A GUIDE TO
MODERN
SCIENCE

Science and Technology in Today's World

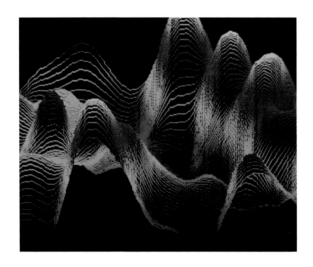

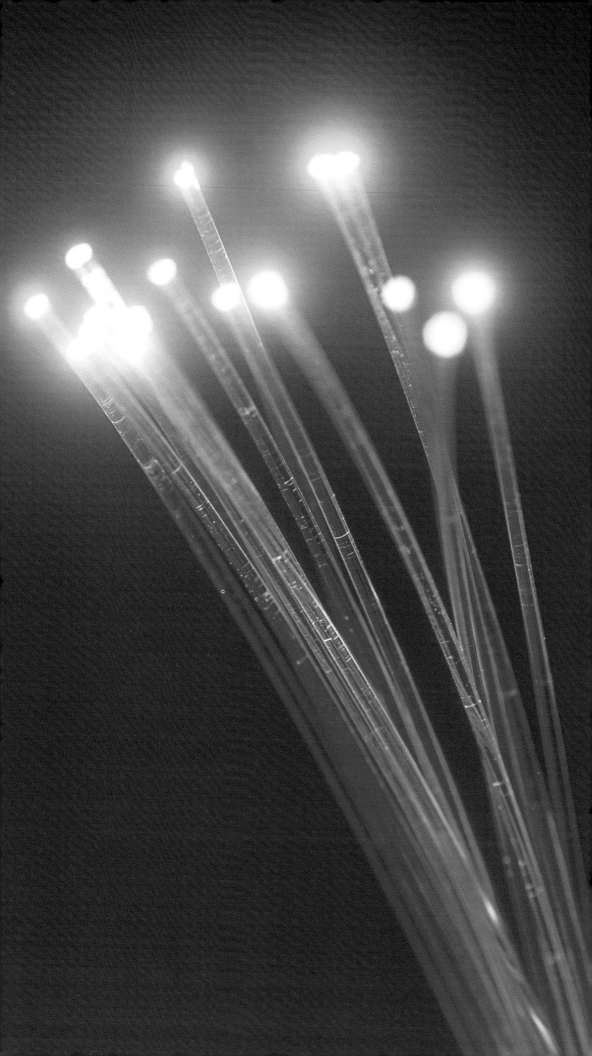

A GUIDE TO
MODERN
SCIENCE

Science and Technology in Today's World

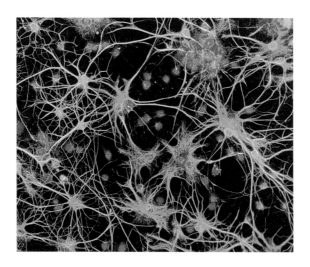

BARRY ANDERSON, STEPHEN BATTERSBY,
LAURIE BECKELMAN, MARCELA BILEK,
MICHAEL BROOKS, JENNY BROWN,
BRUCE BUCKLEY, ROBERT BURNHAM,
ALF CONLON, BEN CRYSTALL, LEIGH DAYTON,
MARTIJN DE STERKE, KAREN MCGHEE,
DAVID MCKENZIE, JULIAN MALNIC,
NATASHA MITCHELL, GRAHAM PHILLIPS,
ABBIE THOMAS, TIM THWAITES, PAUL WILLIS

CONSULTANT EDITOR
WILSON DA SILVA

FOG CITY PRESS

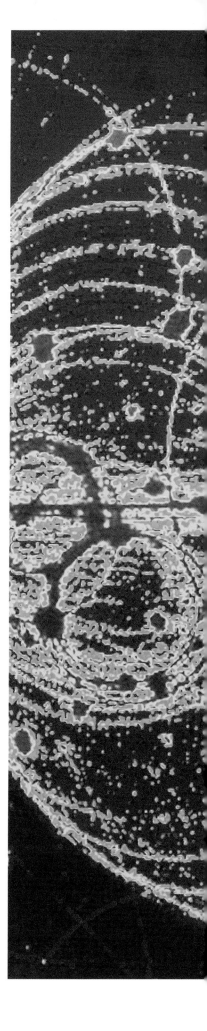

Published by Fog City Press
814 Montgomery Street
San Francisco, CA 94133 USA

Copyright © 2002 Weldon Owen Pty Ltd

CHIEF EXECUTIVE OFFICER John Owen
PRESIDENT Terry Newell
PUBLISHER Lynn Humphries
MANAGING EDITOR Janine Flew
ART DIRECTOR Kylie Mulquin
EDITORIAL COORDINATORS Tracey Gibson, Kiren Thandi
PRODUCTION MANAGER Caroline Webber
PRODUCTION COORDINATOR James Blackman
BUSINESS MANAGER Emily Jahn
VICE PRESIDENT INTERNATIONAL SALES Stuart Laurence
EUROPEAN SALES DIRECTOR Vanessa Mori

PROJECT EDITOR Stephanie Goodwin
PROJECT ART DIRECTOR Sue Burk
PROJECT DESIGNERS Avril Makula, Kylie Mulquin
PICTURE RESEARCH Joanne Holliman

ISBN 1 877019 05 4

Color reproduction by Colourscan Co Pte Ltd
Printed by Kyodo Printing Co (S'pore) Pte Ltd
Printed in Singapore

A Weldon Owen Production

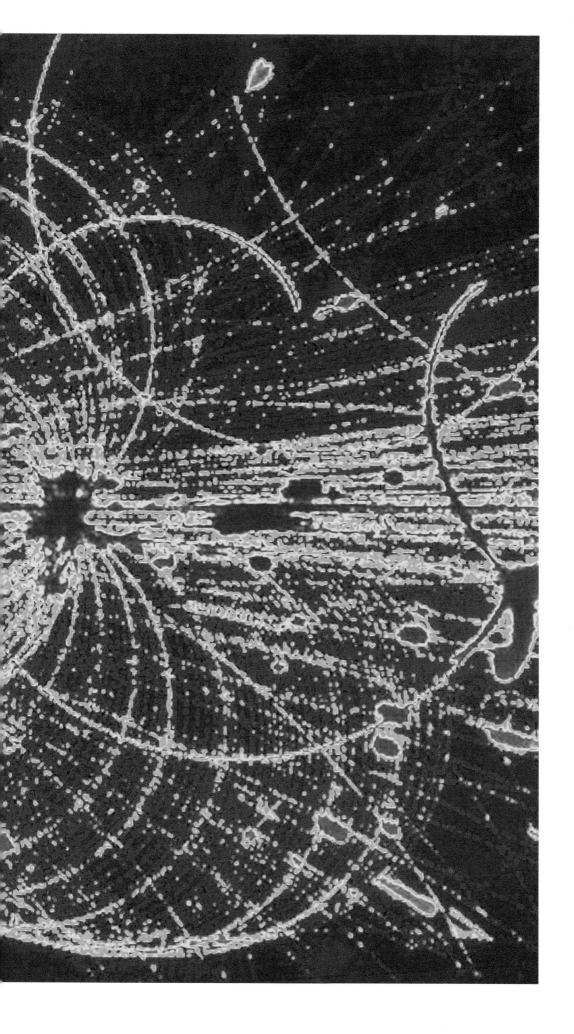

There is a single light of science, and to brighten it anywhere is to brighten it everywhere.

ISAAC ASIMOV (1920–1992), Russian-born American writer

CONTENTS

FOREWORD

This book is by no means an exhaustive list of what the 21st century will bring, nor a detailed road map to the shape of things to come. It is a series of signposts to the future—or perhaps postcards from a future that might emerge based on what we know today.

There is no consensus view among futurists, and certainly none among scientists, technologists, artists, politicians, and business leaders about what this century will bring. But we do know one thing: It will be quite a ride.

Already, some of the things we are able to anticipate are extraordinary enough—imagine those we can't anticipate! But science is a gradual process and tomorrow's leaps will rely on the scientific groundwork that came before. The great physicist Sir Isaac Newton recognized this. When hailed for his discovery of gravity and his laws of motion—still used today by everyone from engineers to airline pilots—he said: "If I have seen further, it is by standing on ye shoulders of giants." No doubt the great pioneers of the 21st century will also climb atop the shoulders of their predecessors, relying on advances and technologies that came before them to push back the frontiers of knowledge.

How does one catalog this in a book? Rather than go through each scientific discipline and make a prediction, we have segmented modern life, and modern science, into conceptual parcels that can be digested separately. At times there is overlap because a problem may have multiple or common solutions. But each potential advance is seen in the context of the chapter focus. Hence, cloning may have huge implications for human medicine, but its horizons stretch much farther—something that is dealt with in the technology section. In the end, there can only be one certainty: We will be amazed.

WILSON DA SILVA
Consultant Editor

INTRODUCTION

Year after year, scientific theories are reviewed and revised, technology is updated, and groundbreaking advances are made that touch our lives in a multitude of ways. Keeping up to date with these changes is no easy feat. *A Guide to Modern Science* helps to clarify what has occurred and explores what might, realistically, occur in the course of the current century. It does not pretend to provide an exhaustive list of how science and technology may contribute to the years ahead, but serves to highlight the rate of substantive progress in many pivotal areas.

The book comprises seven chapters, each of which deals with subjects that are familiar and accessible to every one of us. The subject areas are: The Body, Society, Frontiers of Technology, Transportation, Infotech, The Living World, and Astronomy and Space Exploration. Each chapter contains essays that focus on the speculative achievements of the 21st century, and provide historical background that highlights the cumulative advances of science to date.

Written by a multinational team of scholars and scientists, this book aims to be a provocative and at times sobering look at what the 21st century has to offer in terms of the evolution of science and its place within human culture.

A Guide to Modern Science is part of a series of books that focuses on popular science and natural history. This book and its companion volumes take the reader on a journey of discovery into many intriguing aspects of our world.

The Editors

Introduction

Prediction is a hazardous exercise, particularly in science.
And yet, no century in human history has been so discussed,
debated, and dreamed about, nor so anticipated in film
and literature, as the 21st century.

Part of this is, no doubt, due to the mystique that surrounds millenniums. The 21st century is, after all, the first marker on the road into the third millennium. But there is also something vaguely profound about the number 21. In a number of cultures, it is considered a marker of adulthood and, perhaps subconsciously, we see it as the beginning of humanity's long march toward adulthood after what has been, in the past few centuries, an at times troublesome adolescence. True, there have been wars and hunger, pestilence and holocaust; but there have also been significant advances, uplifting achievements, and moments of great humanity that have made us all proud.

And so it may well be that, more than 10,000 years after the invention of agriculture, 3,000 years since the rise of Ancient Greece and almost 250 years after the arrival of the Industrial Revolution—humans are on the cusp of maturity as a species.

What will the 21st century bring? We only have to look at the 20th century, a period like no other in human history, to get a glimpse.

Never in human history has technology advanced so rapidly, had such an impact on the world around us, and on our ordinary lives. I think of my grandmother, born in the dying years of the 19th century. When she entered the world, the idea of human flight was fanciful. Yet, by 1903, Wilbur and Orville Wright had flown their little engine-powered glider 852 feet (more than half a mile) over the sand dunes of Ohio in the United States. Within a generation, there were regular airline services in Europe and Australia. A generation later, the sound barrier had been broken in the United States. Merely a decade after, the first artificial satellite, Russia's Sputnik, was orbiting the Earth, and it was just over a decade later that two men were walking on the Moon.

To my grandmother, this must have all seemed too much. On a clear and warm night, looking up at the beauty of a full moon rising, she confided to me, "You know, they didn't really go to the Moon." I was puzzled. "You mean, the astronauts?" I asked. She nodded her head. "It was done in a film studio. It had to be. How could someone go to the Moon and back?"

Although now dearly departed, her observation has stuck with me ever since. It wasn't that she believed in conspiracy theories. She just had difficulty accepting such a monumental change in human transport capability could take place in her lifetime. More than 30 years ago, Alvin Toffler coined a name for this condition: "future shock." It's what happens when the familiar becomes strange, changing to the point of being almost unrecognizable—

MUCH LIKE AN ICEBERG, *the majority of what there is to know about the world remains hidden from view. Only half a century ago, Harold S. Osborne unveiled his model for a video telephone (photographed, left), then at the cutting edge of technology. But year on year, science and technology advances, cumulatively, and we can be certain of only one thing: That the process of accumulation won't stop with our generation.*

ONE OF THE GREATEST SCIENTIFIC *and technological achievements of the 20th century was putting men on the Moon (top). But not all great ideas are quite so successful, and it's important to remember that although many technological and scientific marvels may come our way in the 21st century they may not be the ones foretold. Whatever happened to 20th-century jetpacks, the 20-hour working week, and the flying "jet cars" of the futuristic cityscape above, for instance?*

like your street, your town, your bank. When the strange becomes ordinary, like computers that talk to you on the telephone. And when familiar technology you take for granted, technology that act like markers in the map of modern life—cars, telephones, books—when these change drastically, when they are suddenly replaced by strange and unfamiliar technologies, you understandably feel a sense of loss. Often, you don't realize it until it's gone: Some now-redundant technology once meant something to you, and in the back of your mind, you never really expected it to disappear.

It may seem strange, but I feel that way about television. Growing up as I did with a TV set in the house, I couldn't possibly imagine a world without it. Our conversations at school revolved around the things we had in common: Team sports that we played, students and teachers we knew … and what had been on TV the afternoon before. Television shows were our common lodestone, a way we could share experiences not only with each other, but with children in other schools, since TV was a common reference point. But not any more: TV today is a fragmented medium, with a plethora of channels offering a smorgasbord of content. Sitcoms from the 1960s are a click away from the latest computer-animated "live action" battle scenes set in the far future.

A documentary on the nesting habits of frigate birds in Ecuador, shot on color-saturated 16 mm film some 30 years ago, co-exists with a soccer game played only hours earlier in northern England. There is no common reference point anymore … only an avalanche of programming that throws past, present, and, thanks to the popularity of science-fiction shows, even the future into the mix.

We are likely to face many examples of future shock as the years of the 21st century roll past. For this will be a century when technological developments will accelerate to almost breakneck speed, the kind of speed that even those accustomed to the hurly-burly of the late 20th century may find a bit sudden. Like my dear grandmother, I and many of my generation may well find our come-uppance— ourselves a little lost for words in the face of the gigantic steps we as a species are about to take. For humans are standing on a number of unprecedented thresholds: Manipulating the genetic codebook of life, building machines at the atomic level, and living beyond Mother Earth.

One of these milestones we actually passed in the dying days of the year 2000: In the early hours of October 31, a Russian Soyuz rocket blasted off from Baikonur Cosmodrome in the former Soviet republic of Kazakhstan. Aboard were three men: Bill Shepherd, Yuri Gidzenko, and Sergei Krikalev. They docked with the International Space Station on November 2, the first of many crews that will visit the station in years to come. It was a significant event, for on that seemingly innocuous day the human species had taken its first permanent foothold off Earth. From that day forward, humans could no longer be said to only inhabit the surface of their home world—they now had a permanent beachhead to the stars.

There will be many milestones in the 21st century—some of them heralded, some of them seemingly creeping up on us. At times slowly, at times suddenly, our lives will change.

Luckily, we have been well prepared. Not a year goes past where a major movie somewhere in the world does not give us a vision of what life might be like in the near, or even the very far future. Their predictive capacity is questionable, but in their flights of fancy they prepare us for what is to come. Their scenarios may be dystopian, such as the genetically engineered workers who run amok in *Bladerunner* (1982), or hopelessly optimistic, such as the adaptation of the H. G. Wells novel, *The Shape of Things to Come* (1936), where a group of scientists re-build civilization shattered by nuclear war into an ordered and just society guided by scientific principles. These and many films like them over the past century have tried to encapsulate a vision of what the future might hold for us. But in the end, they tend to be a projection of fashions and concerns at the time they were made: *Things to Come* represented a fear of impending war but a determined belief in the ability of science and technology to overcome. *Bladerunner* depicts Los Angeles circa 2019 where Japanese culture has subsumed the mainstream, surveillance technology pervades everything, and genetic engineering is run by greedy corporations disinterested in ethics.

Some of these themes echo true. As you will read in this book, unraveling the genome will give us unprecedented power over the

ON DECEMBER 30, 2000, THE WORLD CELEBRATED
the birth of a new millennium and possibly the most anticipated century ever, a century in which technological and scientific breakthroughs are expected to accelerate at almost breakneck speed, changing our lives beyond recognition.

15

THE DISCOVERY OF THE QUARK *(above) is one of the many scientific successes of the 20th century. The more we discover, the more apparent the interconnectedness of life becomes to us and we look to sustainable alternatives such as the breeding and farming of fish in aquaculture farms (left).*

biological processes in our bodies; to halt disease, correct genetic disorders, improve our chances of a longer life, and even make some improvements here and there to our physical bodies. We will have the capability of making life appreciably better for all. It is also likely that this century, we will eventually hold the keys to the most powerful instrument ever created—the genetic book of life—and be able to tinker with it as we wish.

However, with this must come great wisdom, for as we have seen in the past century, in the headlong rush of enthusiasm to exploit such great advances, we can occasionally blunder disastrously. But these advances also present us with exciting opportunities to improve the lives, the health, and the wellbeing of millions—as well as saving our environment and the fate of countless other creatures.

And herein, perhaps, will lie one of the major differences between the 20th century and the 21st: An appreciation that we live in a closed system. That like the astronauts living aboard the space station high above us, we too live in a spaceship, albeit a much larger one, called Earth. At no time in history have humans been so capable of altering their environment so dramatically—from the detonation of nuclear weapons to the rapid destruction of forests, from fishing species to extinction to melting the polar icecaps. We enter the 21st century much more aware of the interconnectedness of all life on Earth.

It's a philosophy that pervades many of the branches of modern science, and is reflected in the solutions that scientists are now seeking to everyday problems and long-term dilemmas. They look for "sustainable" solutions, technological advances that are not just fixes, but actually work in harmony with the architecture of Spaceship Earth. This is not just the idle philosophizing of a scientific or fashionable elite: It's an approach that is winning support in boardrooms across the planet and has already changed the course of governments—local, state, and national. As the cost savings from the reduction of waste and the benefits to lifestyle and the economy become more tangible, we will see an explosion of this approach … to the point where we will wonder how we ever did it any other way. This will be one of the more pleasant "future shocks" to come, and perhaps one of the best signs of our growing maturity as a species.

But beware: Not all that we dream, hope, and anticipate, nor all that we fear, will necessarily come about. At times the greatest ideas and the most marvellous inventions can founder. Whatever happened to videophones, jetpacks, and walking, talking robots? I recall, during my visits to the local library, reading with anticipation of these and other wonderful developments the 21st century would bring; so excited was I as a child that in the evenings I dreamed of flying about with jetpacks strapped to my back. Such advances seemed quite reasonable to me at the time. But others, such as the idea that computers—monstrous and noisy machines with spinning tapes in my time— could one day be carried in your top pocket,

THE WORTH OF ANY TECHNOLOGY *is not always immediately clear. Personal computers were looked upon as curiosities for years until the first spreadsheet programs were devised and they became useful as business tools.*

sounded a little far-fetched. Alas, just decades later, I walk around with a pocket digital assistant that carries one million times the on-board memory power used on the computer aboard *Eagle,* the lunar excursion module that landed men on the Moon.

Nobody can truly predict what the future will bring. As a journalist and editor, I have covered developments in science and technology for more than 15 years, and I still find myself surprised and occasionally astounded by the unexpected twists and turns. It's what makes science such a fascinating branch of human endeavor.

Each year, we know a little more about the world, and a little bit more about this immense universe around us. Science is like a candle held against the dark; each year, another candle is lit, and our surroundings become a little less dark. But the cave of knowledge is a vast expanse that seems to know no bounds. We will never be able to light the whole cavern.

While there is no way to know what path the future will take, you can at least arm yourself with a good map. This book represents the world's best thinking on various topics, encapsulating the reflections of the best minds and canvassing possibilities our parents would never have dreamed of. It's the kind of reading my grandmother could have done with, so that she wouldn't feel so surprised that so much had changed around her, so astounded by concepts she had never considered that she would steadfastly decline to accept them.

And, for the record, men really did walk on the Moon.

WILSON DA SILVA, *Consultant Editor*

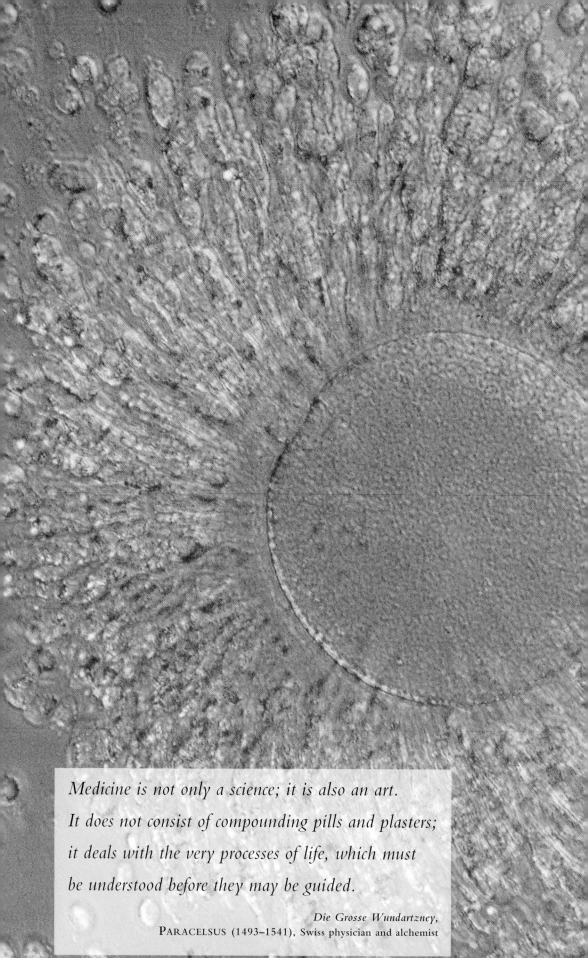

Medicine is not only a science; it is also an art.
It does not consist of compounding pills and plasters;
it deals with the very processes of life, which must
be understood before they may be guided.

Die Grosse Wundartzney,
PARACELSUS (1493–1541), Swiss physician and alchemist

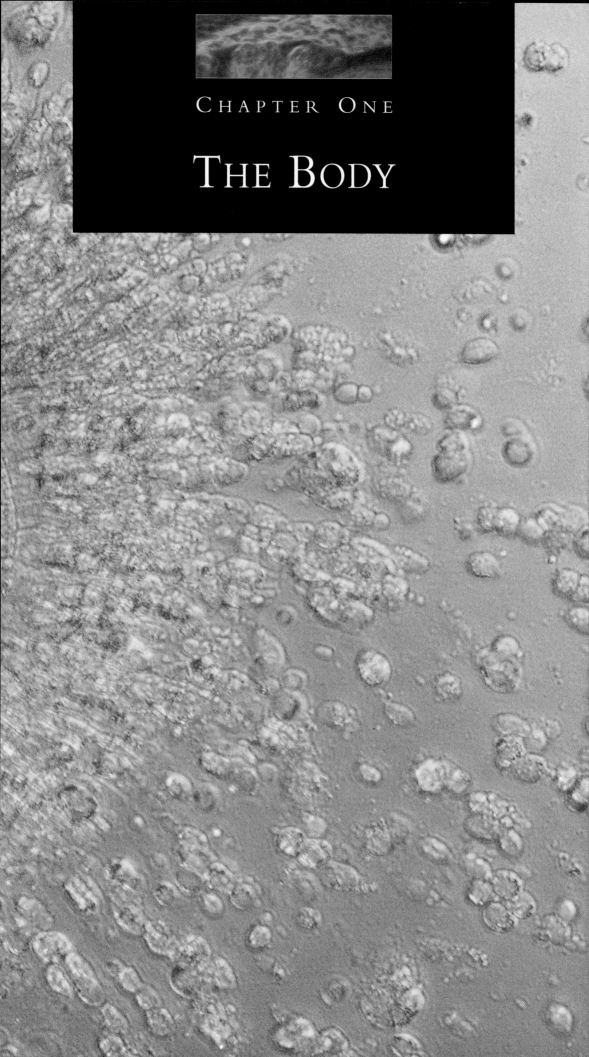

CHAPTER ONE

THE BODY

A Cure of One's Own

*In the future you may have your own medical
report card detailing your risks for future diseases
based on the genes you have inherited.*

Nature needed but four letters to pen the manual of life. Those letters—A, C, T, and G—known as nucleotides, represent the four chemical bases that form the genes of all living organisms, from the humble bacterium to the sophisticated human being.

A significant number of those genes are the same across species. We share 75 percent of our genes with the mouse and 98.4 percent with the chimpanzee, our closest relative. We share still more with one another. Genetically speaking, you and every other human being are 99.9 percent identical. But what a difference that remaining tenth of a percent makes. Among other things, it shapes both your susceptibility to disease and your response to treatment.

Researchers are mapping the genetic differences that predispose people to disease and influence how they respond to medication. Once they are known, these differences will lead to individualized treatments that fit as well as a Hong Kong suit. At some distant date, you may carry your genetic profile on a disc. When slipped into your doctor's computer, your disc will reveal the genetic quirks that will make you respond well to one drug but not another. It will unveil the health vulnerabilities hidden in your genes. Are you at risk for depression? Alzheimer's disease? stroke? Your DNA disc will tell all, enabling your physician to suggest any appropriate monitoring and preventive care.

DNA discs won't be as ubiquitous as a driver's license anytime soon: Scanning all your genes today could cost you $20,000 and take up to two weeks. Faster and cheaper technologies are on the way, though, and large-scale genetic screening will become feasible. Whether it will be useful will depend on the ability of researchers to solve an even greater problem, and that is to determine what all those genes do.

MYSTERY GENES

The scientific gold rush that produced a rough draft of the human genome at the dawn of this century was but the first step toward developing a rational system of medicine based on genetic differences. Scientists still have to fine-tune the gene map, filling in the missing data and correcting errors that the rough draft contains. They hope to do so by 2003. At that point, they

KEEPING FIT *will be important even in an era of
custom-made medicines. Environment, as well as
genes, will always have an influence on health.*

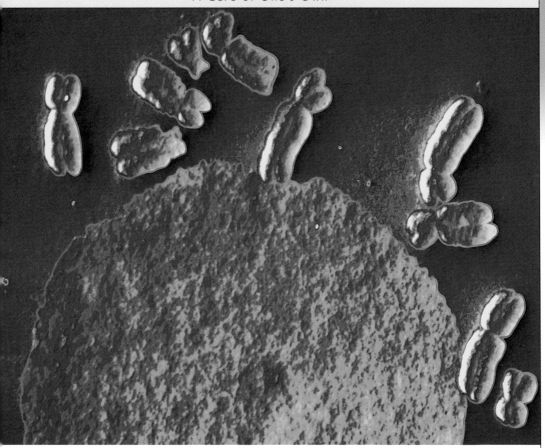

will have a list of all the genes that make up a human, and they'll also know where each one lies on the 23 paired chromosomes that carry our genetic code. But they still won't know what most of those 40,000–70,000 or so genes actually do. To transform personalized medicine from dream to reality, they will need to identify the genes' functions and carefully map the complex interactions between them that could possibly give rise to disease.

The race to do so is on, as is the all-important push to identify the genetic mutations that are meaningful for human health. We already know some of these, of course. Genetic mutations are well-recognized culprits in inherited illnesses ranging from cystic fibrosis to Huntington's chorea, and researchers have had mug shots of many of them for years. But the vast majority of genetic variations have more subtle effects on health. These differences, which are known as single nucleotide polymorphisms (SNPs, pronounced "snips"), are key to the personalized medicine of the future.

ONE IN A THOUSAND

SNPs make a significant contribution to human diversity. Every 1000th letter in your personal genetic manual is different from most other people's, but it is still not unique. As little as one percent and as much as 49 percent of the

CHROMOSOMES, *thread-like structures occurring in every cell nucleus, condense and replicate during cell division (above). Chromosomes carry a person's genes, the inherited instructions for life. Genes influence our responses to medical treatments.*

THE NEW ERA *of molecular genetics can be dated back to 1953, the year in which James Watson (above, left) and Francis Crick (above, right) worked out the structure of DNA. They are pictured here with their model of part of a DNA molecule.*

21

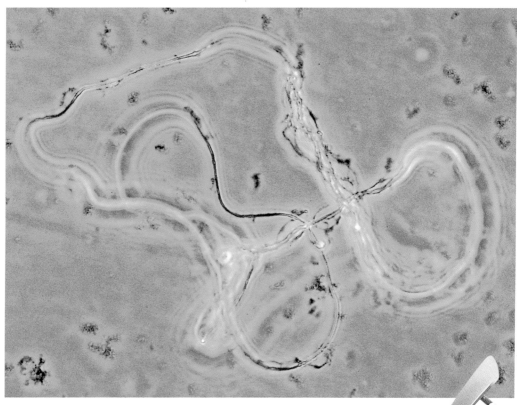

population shares each variant with you. These one-in-a-thousand SNPs are the simplest of genetic variations. Each is a difference in just a single nucleotide pair within a gene. The nucleotides are the four base chemicals—A, C, T, and G—which form the rungs of the DNA ladder. The order in which these letters appear determines what a gene does, or, to be more precise, what proteins it produces.

Nearly all genes provide instructions for manufacturing proteins, the chemicals that form most of the human body and do its work. Change one of a gene's nucleotides and you change the proteins the gene produces. This in turn changes the way the cell—and the body— functions. That's why SNPs are so important. One SNP can make the difference between a protein that amplifies a drug's effects and one that diminishes them. Several together may predispose someone to complex diseases such as cancer or schizophrenia.

Even SNPs with no direct effect on health may be medically significant. By definition, each SNP occurs in at least one percent of the population. Mutations are far more rare. A harmless SNP that occurs close to a deleterious mutation could serve as a marker, pointing to the otherwise difficult-to-detect troublemaker.

A FIELD IS BORN

An ambitious effort to identify and map a significant proportion of all human SNPs is under

ONE DAY, *doctors will be able to "read" health vulnerabilities buried in your DNA to determine the best medicine for you. The above micrograph shows thread- like strands of human DNA.*

THE LONG AND SPIRALING *strands of a DNA molecule consist of a sugar-phosphate backbone to which amino acid bases, which are called nucleotides, attach. Each base links to a complementary one on the other spiral, forming the rungs of the double helix (right).*

0.34 nm

3.4 nm

Chromosomes

IDENTIFYING SNPs *could lead to more effective and safer treatments. Having identified the SNP that turns a potential treatment for childhood leukemia life-threatening, scientists can now screen the patient before adjusting the dosage accordingly.*

way and is already yielding far better approaches to treatment. For example, researchers have identified many SNPs that influence the cascade of chemical events necessary for the body to absorb, transport, break down, and eliminate drugs. One of these SNPs will turn the usual dose of a potentially life-saving treatment for childhood leukemia into a life-threatening one. A screening test now available enables a doctor to easily identify any children with this particular SNP and make the necessary adjustment to their dosages.

Hopes of developing many more tests like this one, as well as improved medicines, have spawned the new field of pharmacogenomics— the development of drugs tailored to specific gene variations. Researchers estimate that many of today's drugs work only 30 to 50 percent of the time. More targeted treatments hold the promise of greater efficacy, fewer side effects, and a vast reduction in the adverse reactions that claim more than 100,000 lives a year in the United States alone. Such fine-tuned pharmaceuticals will be the first successes of pharmacogenomics.

How soon the field will provide new treatments for currently incurable diseases is much more difficult to say. Before this century is out of its 20s, scientists expect to discover the genes implicated in all known human diseases. Some of these discoveries will undoubtedly yield the wonder drugs of tomorrow. But it is inevitable that others will disappoint. The case of cystic fibrosis is a sobering reminder of just how hard drug development can be. Scientists have long known the location and the sequence of the single gene that causes this life-threatening disease, yet to date they have been unable to develop a cure for it.

On a more positive note, however, the massive investment made in improved technologies, not only for identifying

genes but also for deciphering their functions and variations, holds out much more hope for the future than ever before.

TROUBLING QUESTIONS

Like any medical approach that lays bare the genetic map of the self, pharmacogenomics raises ethical concerns. Might someone who does not respond to the best treatment for a costly disease be denied health coverage? Might developing drugs for people with rare genetic variants be prohibitively expensive, leaving them without potentially life-saving remedies available to the majority? Who should have access to a person's genetic profile? A pharmaceutical company eager to sell its wares? An employer? An insurance company? Society, not science, will need to answer such questions.

DECODING THE BOOK OF LIFE

Scientists around the world—and anyone else who is interested— can now read the book of life online. That book, the human genome, is the full set of blueprints for a human being. It's not yet complete, but even its rough draft is transforming our current understanding of human biology. The genome map is the product of a multinational effort

called the Human Genome Project led by Francis Collins (above, left), as well as of a parallel effort by private company Celera Genomics in the United States, co-founded by Craig Venter (above, right).

In 1990, the Human Genome Project began to develop the technologies and techniques needed to sequence all of the genes in a human body. Scientists from 16 institutions in six countries went on to map the genome. Their efforts led to the discovery of thousands of genes coding for even more proteins, the building blocks of life.

Once scientists identify the proteins a gene codes for, they can work out what it does and how to manipulate it to enhance health. This is no easy task. To succeed, it will take an effort as intensive, and new technologies as powerful, as those of the Human Genome Project. But the potential payoff is a future in which more precise treatments prevent, abate, or cure diseases that cause suffering today.

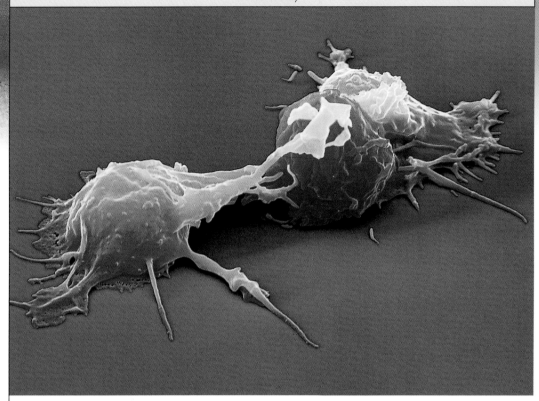

Gene Therapy

Gene therapy could be to this century what antibiotics

were to the last. It promises enhanced treatment—perhaps

even cures—for a staggering number of diseases.

The novel technology of gene therapy could vanquish viruses and inherited illnesses, slow the growth of tumors, and halt progressive diseases such as Parkinson's or arthritis. The concept behind it is simple: If a gene doesn't work properly, then just replace it with one that does.

Researchers now believe that our genes play a role in most—if not all—human illnesses. Some genes are faulty from birth and destine their owners to disease. Others predispose. Still others succumb to the assault of environmental toxins or viral coups, giving rise to cancers or to diseases such as AIDS. By developing the means to replace defective genes or at least block their actions, researchers hope that they can stop their deleterious effects.

Though the concept of gene therapy may seem simple enough, its execution is anything but. Scientists must first get genes into the right cells in sufficient quantities to affect health. Then they must get the genes to work in the way that they should. What's more, they have to set this complex armory in place without sounding the body's sensitive "invader!" alarm.

Gene Ferries

Of these formidable challenges, the greatest is developing successful vectors, the ferries that transport genes to their intended targets. Most vectors are viruses, selected because they are masters at invading human cells. Genetic engineers remove the virus's disease-causing genes, splice in the corrective human one, and then unleash the modified virus in the patient's body. The virus weaves its way into cells, where the new genes begin functioning.

All viral vectors used to date have limitations, however. Some vectors invade only dividing cells and are thus useless against noncancerous diseases of the brain, liver, or lungs. Some infect too widely, delivering their cargo where it isn't needed; others are too small to carry large human genes. And of course, the body is primed to attack viruses. An immune response can decimate the vectors, rendering

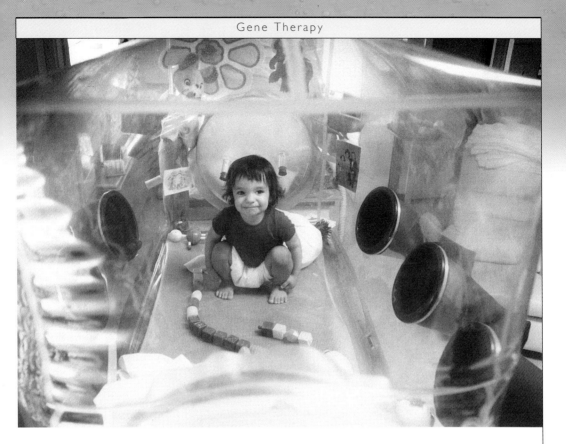

DELIVERING CORRECTIVE genes to damaged cells would ensure an effective attack of human natural killer cells on a cancer cell (opposite). Gene therapy has already given infants born without immune systems the chance to live outside a bubble like the one occupied by David in the 1970s (above).

treatment ineffective. Even worse, it can kill. Scientists are continuing to modify viruses to prevent such problems. They are also designing non-viral vectors and alternative methods of manipulating genes to protect health. The most promising of these seeks simply to stop faulty genes from expressing themselves, rather than replacing the genes entirely.

New Hopes

In 2000, a decade after the first gene-therapy trials in humans, scientists reported the field's first unquestionable success. Researchers in France treated three infants born with incomplete immune systems, a deadly condition that sentences such children to life in a germ-free bubble. Ten months after gene therapy, all three of the babies' immune systems were normal. Although researchers cautioned that only time would tell if they would remain so, for the moment at least gene therapy

had freed these children to hug their parents, pet their dogs, and poke around sandboxes without fear of dying.

The millennium brought news of other gene-therapy successes, too. Experimental treatments revived dead heart tissue, shrank tumors in patients with advanced melanoma, and enabled three patients with hemophilia B to reduce their reliance on infusions of a blood-clotting factor their bodies can't produce. These promising results bode well for the future of a field that is still in its infancy.

CHANGING GENES INSIDE THE WOMB

A needle pierces the amniotic sac to extract fluid for genetic testing. Days later, another one gently pricks the fetus itself, delivering genes to reverse a detected defect. This vision of curing diseases in the womb is as contentious as it is tantalizing. Many researchers believe that fetuses are ideal candidates for gene therapy. Their immature immune systems are less likely to attack viruses used to deliver genes, and their rapidly dividing cells are more likely to incorporate them. Concerns abound, however, ranging from worries that unethical geneticists will indulge parents' requests for "designer babies" to fears of unanticipated consequences of the therapy itself. The current technologies can't assure that the genes enter only the targeted cells, for example. It's possible that they could infiltrate others, causing harmful mutations. Critics are most concerned that the new genes might lodge in reproductive cells (called germ cells), creating changes that would reverberate through generations. The prospect of correcting genetic defects before they cause damage is so compelling, however, that research is very likely to progress further.

The Kindest Cut

Surgery without scalpels will become the norm as researchers develop increasingly sophisticated ways to reach and see inside the body without opening it up.

In the surgical suite of the future, the kindest cut may be no cut at all. The lesions that wreak havoc on human health can lurk beneath the brain's deep furrows or against an artery's wall. They can hide in the heart's throbbing muscle or along the twisting tunnels of the gut. To see, reach, and treat them, surgeons have long needed to peel open the body. But today's researchers are developing new ways to diagnose and treat old ills without the trauma and long recovery times synonymous with traditional surgery.

These new approaches substitute lasers and shock waves for scalpels and saws. They augment the surgeon's vision with 3-D images broadcast from the body's depths. They replace incisions large enough to accommodate a surgeon's hands with small slits through which slender robot fingers operate. Collectively called minimally invasive therapies, these new technologies will shape future medical care.

LESS IS MORE

The quest to peer inside and fix the human body without laying it open is nothing new: Hippocrates examined hemorrhoids through a primitive scope and described noninvasive treatments for them and other conditions more than 2,000 years ago. Not until the closing decades of the 20th century did technological advances make a future without open surgery feasible, however. Powerful new imaging, energy, and computer technologies, along with ever-smaller and more precise surgical tools, ushered in the era of less-is-more medicine.

Miniaturized cameras gave doctors inner vision as they snaked tool-carrying catheters through the body's bloodstream to unclog an artery or block off a bulging and dangerously thin piece of arterial wall. Fiber-optic endo-

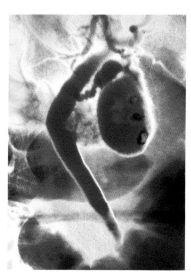

A FALSE COLOR X-RAY *of a gall bladder shows gall stones (colored green). The technique used to obtain this image, endoscopic cholangiography, involves injecting a contrast medium into the bile duct via an endoscope.*

scopes carried a light, surgical tools, and a camera into the body's hollows, where surgeons could then biopsy a suspect lesion, stop bleeding, or remove diseased tissues. Magnetic resonance imaging, ultrasound, and other imaging technologies projected pictures of the body's organs to aid diagnosis and guide surgery.

Refinements of these technologies will enable tomorrow's doctors to perform increasingly complex operations less invasively. Smart catheters will make the bloodstream ever more navigable, enabling access to practically any organ through a tiny puncture hole. They will deliver drugs, genes, and cellular transplants directly to their targets. New or improved imaging technologies will bring the body's organs into ever-sharper focus, making the need for surgical biopsies obsolete. These technologies will let surgeons preview the abnormalities they must remove and will guide surgery itself, generating images that provide more information than the surgeon's own eyes.

Some imaging devices will be as small as multivitamins. Indeed, a camera-in-a-pill has already won United States government approval for use in diagnosing bleeding in hard-to-image sections of the intestines. When swallowed, it beams back photos of the digestive tract. Tomorrow's endoscopes promise to be as revolutionary. Some will be the equivalent of surgical Swiss Army knives. Armed with optical filters, lasers, and tissue markers, for example, they will diagnose and treat diseased tissue in place without disturbing normal cells.

ROBO-DOC

Minimally invasive surgery has transformed procedures such as gall bladder removals and knee repairs, resulting in less pain, fewer complications, and shorter recuperation times for patients. But the long, inflexible instruments used for today's endoscopic surgery cannot perform dexterous motions, such as tying sutures, which many more complicated procedures require. Robot fingers can.

Jointed, flexible, and under a surgeon's remote control, robots have now been used to remove kidneys and prostate glands, to operate on the heart, and to biopsy and destroy breast tumors—all through incisions no bigger than a dime. Robots like that of Dr. Yik San Kwoh (above), shown simulating the insertion of a biopsy needle into a model's head, may soon perform brain surgery as well. Still experimental, they will some day be routine members of the surgical team.

HEALING RAYS

Even the buttonhole incisions of endoscopic surgery may seem barbaric in the future, as researchers harness the power of energy to heal. Energy sources such as lasers, electron beams, or high-intensity ultrasound can pass safely through the body, affecting only targeted tissue. As researchers perfect techniques for reaching deep tissues and precisely targeting the energy beams, noninvasive surgery will increasingly be the norm. Already, shock waves that stimulate a bone's own healing mechanisms are mending recalcitrant fractures, and high-intensity focused ultrasound is obliterating breast tumors. Another new approach, photodynamic therapy, uses a light source such as a laser to activate photosensitive drugs that have infiltrated cells. Upon activation, the drugs kill the cells, making this a promising treatment for cancer.

The ultimate in minimally invasive therapy may not come until MEMS prowl the body, however. Swallowed or injected, these micromachines will use gyros and accelerators to propel themselves toward a trouble spot. Once there, they will diagnose the problem, radio their findings back to a physician, and dispense treatment at the doctor's command, all without causing as much as a burp.

WITH THE ADVENT *of less invasive procedures, compound fractures (below) require surgery but many fractures may not.*

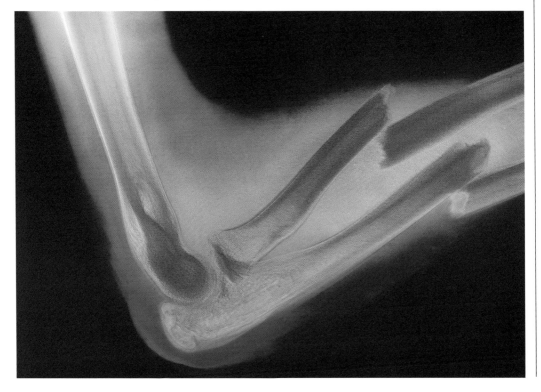

Repairs and Replacements

A deep freezer stocked with human livers, kidneys, and ready-to-beat hearts may sound fantastic, but off-the-shelf organs could be among tomorrow's routine cures.

Within the next decades, scientists hope to grow a human heart. They are already making its parts. Valves, vessels, and patches of heart muscle are growing in labs all over the world. Although these parts can't make a whole heart yet, their creators are confident that one day they will. Participants in the Living Implants from Engineering (LIFE) initiative, these scientists believe that once they can grow a heart, they can eventually grow any organ—even one as complex as the liver. Their goal is to make living replacement body parts as commonplace as penicillin.

The need for such a supply of human tissue is vast. In the United States alone each year, lab-grown hearts and lungs, livers, and kidneys could save the lives of 65,000 desperately ill

TISSUE ENGINEERS *may one day enable people to do what salamanders (right) do naturally—regenerate limbs. The first natural prostheses, however, will be lab-grown.*

patients who anxiously await transplants in the knowledge that the necessary organ might not become available in time. Natural cartilage and bone could put a new spring into the step of 500,000 people who receive joint replacements each year, while tissue-engineered arteries may save 600,000 bypass patients from the added trauma of having vessels snipped from their legs in order to repair their hearts. The thousands of infants born with faulty heart valves, women who lose breasts to cancer, people disfigured in accidents or burned in fires—all could benefit from lab-grown replacement tissues.

CARTILAGE *grown from a patients' own cells will one day make hip implants (below) obsolete; it is already helping some patients for whom conventional knee surgery has failed.*

BEAUTIFUL BATHSHEBA'S *rippled left breast (left) suggests that the model, Rembrandt's mistress Hendrickje, may have had breast cancer. Women of the 21st century may be able to preserve the beauty of their breasts despite mastectomy by opting for implants of natural, lab-grown tissue.*

BODY-BUILDING

The ultimate in organ replacement, however, is tissue engineering. This multidisciplinary field draws on the expertise of cell and molecular biologists, engineers, and specialists in materials science to replicate the body's growing conditions in the lab. Tissue engineers seek to create organs that are indistinguishable from the real thing. The basic materials of these organs are cells, scaffolds, and nutrient-rich growing mediums. Engineers begin

People who face the agonizing wait for new organs today may meet with several life-saving options in the future. Some scientists are developing transgenic animals, new breeds of pigs or sheep that could possibly serve as organ donors for humans. Others are combining living cells with mechanical devices to create stopgap substitutes for livers or kidneys. Bioartificial livers, for example, might work in the same way as kidney dialysis machines but would use living liver cells to filter wastes from the blood and to supply the proteins and other chemicals normal livers provide. Such devices could extend life for patients awaiting transplants.

work by harvesting the requisite cells. They grow these cells in a culture to acquire a sufficient supply, then seed the cells onto a biodegradable scaffold shaped like the desired organ. More than a tailor's dummy, the scaffold is typically impregnated with growth factors or other chemicals that help direct the tissue's growth. Finally, the tissue engineers place the cell-coated scaffold into a bioreactor that simulates the natural growing conditions of the body. Within the bioreactor, a kind of chicken soup for the cell bathes the growing tissue. Then, over a period of time, the tissue develops and the synthetic scaffold dissolves, leaving

THE IMMORTALS

A cure for diabetes may be growing in a lab in the United States. Researchers at the School of Medicine, University of California in San Diego, have cultured an immortal line of human beta cells, the insulin-secreting cells of the pancreas. Transplanted into mice, the cells function properly. Human trials are some years away, but if these cells can be grown in sufficient quantities to treat people, they could provide a permanent source of insulin for sufferers of diabetes.

A FETAL PIG'S PANCREAS *(right) is a promising new source of insulin cells for diabetics. But first, scientists must resolve the technical and ethical issues of xenotransplantation, the use of tissue from one species in another species.*

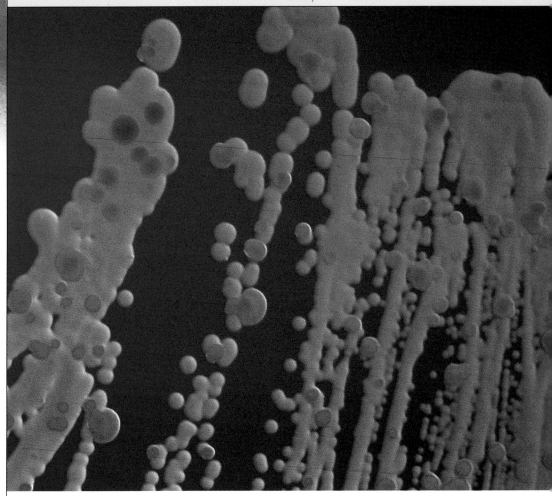

behind a heart valve or an artery, a nugget
of cartilage or bone.

Generating human body parts isn't quite
as straightforward as this recipe suggests,
however. The precise growing conditions
within nature's own bioreactor—the body—
are not wholly understood. Cells, like the
being they constitute, develop in response
to both genes and environment. To take but
one example, cells get developmental cues
from their neighbors and can behave strangely
when isolated. If you grow smooth muscle
cells on their own, for instance, they often
stop contracting and start secreting collagen,
which then forms scar tissue. The muscle cells
develop properly, however, when coupled
with the endothelial cells that would normally
line an artery or a hollow organ such as the
bladder. Scientists aren't yet certain why, but
they suspect that communication between the
two types of cells is key.

NEO-ORGANS

Capitalizing on this cellular harmony, Harvard
Medical School researchers in the United States
seeded smooth muscle cells on the outside of a
scaffold and endothelial cells on the inside to

GROWING CELLS IN CULTURE (above) is the first
step in creating human tissues. The cells are then seeded
onto a biodegradable scaffold before being placed inside
a bioreactor that simulates the body's natural growth conditions.

grow the first fully functional internal organ—
a dog's bladder. They transplanted six of these
neo-organs, as tissue-engineered versions of the
real thing are called, into dogs. Within just
three months, the dogs' bodies had supplied
the new organs with blood vessels and nerves,
producing bladders that held and expelled
urine as reliably as their natural counterparts.
Since then, the researchers have grown human
bladders, making it very likely these will be the
first tissue-engineered internal organs to be
tested in people.

Growing meaty organs such as the liver or
the heart poses a greater challenge. The tatting
of arteries, capillaries, and veins within these
organs assures that nutrients and oxygen reach
every cell. The researchers must create an
analogous network of vessels to develop thick
tissues in the laboratory. Without it, the
growing medium cannot penetrate the developing organ, and cells will starve and begin to
die. Researchers are testing approaches to this

THE POWER OF THOUGHT

Nerve cell implants to reverse paralysis and movement disorders are likely to become a major new treatment, but they aren't the only hope for patients with these conditions. Plastic muscles that bend in response to an electrical signal are in development, as are devices that might enable a patient to move a prosthetic limb by sheer thought. Researchers at the Massachusetts Institute of Technology and Duke University recorded the brain activity of an owl monkey (like those at right) as it learned to move an object. They then developed a computer program that isolated the brain signals that instructed the monkey's arm to move. As the computer "read" the monkey's intent, it fired instructions to a robot arm, which moved in perfect synchrony.

challenge that include spiking the scaffold with a substance that promotes the growth of blood vessels and fabricating scaffolds with vessel-like channels already in place.

At the same time, researchers are patiently deciphering the rules of development, learning, for example, when one signalling molecule bows out and another steps in to guide the growth of an organ with multiple cell types or functions. They are figuring out how to simulate not simply the chemical milieu of the body but also the mechanical forces that strengthen the developing tissues, enabling an artery to withstand the pumping pressure of the heart, or a tendon to accommodate an athlete's leap. They are considering how to store an inventory of organs and contemplating the best source of cells for the replacement of aging body parts. The patient's own cells are ideal because they won't cause rejection, but post-midlife cells don't grow robustly. Even older people, however, have adult stem cells. These immature cells are part of the body's natural repair system and can be coaxed in the lab into forming many types of cells. They, or even the more versatile embryonic stem cells, may provide the solution.

EARLY SUCCESSES

While headlines announcing the first human neo-heart are a decade or more away, some tissue engineering successes are already helping patients. Commercially available engineered skin is helping burn victims and patients with leg ulcers, and implants of cartilage-producing cells harvested from the patient's own body, then multiplied in the lab, are restoring knees. Researchers have grown cartilage shaped like ears and noses too, but such living prosthetics are still in the early stages of development.

The trickle of tissue-engineered products making it into the clinic will become a torrent over the next decades. On their way to creating a human heart, participants in the international LIFE initiative, as well as others, will perfect valves and arteries. Meanwhile, American researchers at the University of Michigan School of Dentistry have engineered skin and gum cells to form bone. Seeded into collagen sponges and then inserted into openings in a rat's skull, the implants nearly closed the openings within a month. Other researchers, at Carolinas Medical School in Charlotte in the United States, have harvested cells from the leg and buttock and grown them into a soft tissue mass that could form a more natural alternative to current breast reconstructions. Still others are growing intestines, kidney cells, ligaments, and corneas. Name a tissue and the chances are that somebody, somewhere, is trying to grow it. And replacement tissues are not the only potential application of tissue engineering. Researchers can insert genes into the cells they grow and thereby transform those cells into delivery systems for gene therapy. In this way, they could one day treat a disease such as diabetes with a mere squirt of new cells.

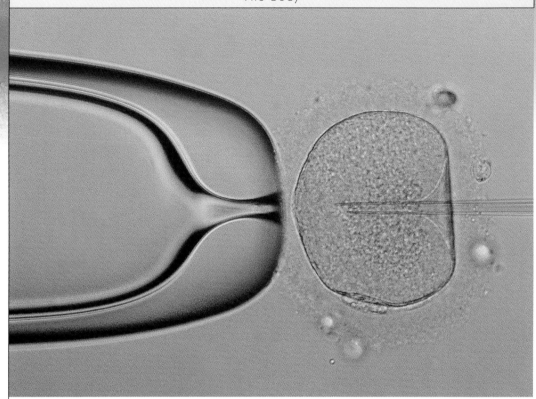

Reproduction and Contraception

The reproductive technology of today, together with genetic engineering, holds the prospect of revolutionizing our species.

Reproductive technology has gone well beyond simply producing a child—a scientific breakthrough achieved back in 1977, when Louise Joy Brown became the first "test tube" baby. Today, in vitro fertilization (IVF) and genetic engineering can be used to choose gender and diagnose genetic defects and diseases in an embryo. Genetic engineering offers the possibility of curing diseases before a child is born and, in the longer term, of changing the genes passed on to the next generation.

But with this new technology come many ethical dilemmas. Once we know an embryo may not be genetically "perfect," do we have the right to decide whether it survives? Should we start picking and choosing the genetic traits that we want in our children? And who can say which are the preferable genes in our species— the ones to keep, and the ones to eliminate?

PRENATAL TESTING AND DIAGNOSIS

The science of detecting genetic abnormalities in human embryos is progressing rapidly. Until recently, the only available tests used chorionic villus sampling or amniocentesis. Both proce-

dures involve extracting fetal cells from the placenta; both carry a small risk of miscarriage.

A new test can analyze fetal cells found in the mother's bloodstream. If a problem is detected, it could be due to the fetus having too many or too few chromosomes, or it could be due to aberrant genes like those for sickle-cell anemia, autism, and cystic fibrosis. (The battery of tests to detect rogue genes is growing daily. It includes a new biochip imprinted with hundreds of DNA probes, which are used to screen embryos before implantation in IVF procedures.) The parents are then left with the unenviable choice of whether to keep the child or to terminate the pregnancy. Often, the decision is influenced by whether the gene will cause immediate serious health problems for the child or whether it will only manifest itself after a relatively long and healthy life.

GENE AND GERMLINE GENE THERAPY

But what if the unborn child could be cured of the genetic disease? This is now becoming a reality through the use of gene therapy, either by replacing defective genes or switching them

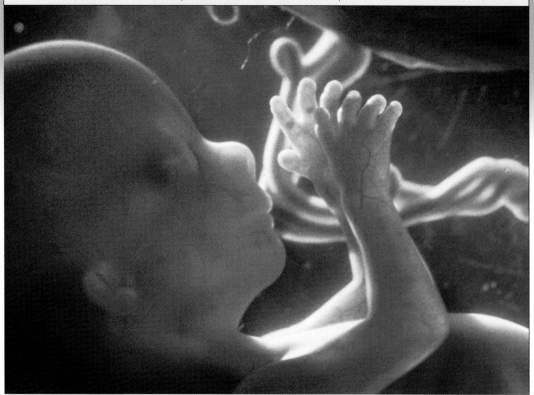

A MICRO-NEEDLE *(at right of the light micrograph above, left) injects the genetic material of a sperm into a human egg during the process of IVF. IVF and genetic engineering are now used to diagnose diseases and genetic defects in embryos. Once diagnosed, it is then technically possible—though it has not been done in practice—to replace the defective genes with healthy ones while the fetus is still in the womb (above).*

off by ferrying a desired gene into the body on the back of a harmless virus, to infect the defective cells. To date, most gene therapy has involved adults. But it could be more effective in a fetus because the immune system of a fetus is less developed and therefore far more accepting of any foreign gene-carrying virus. Gene therapy remains extremely controversial, however, because of the possibility that introduced genes may be passed down from generation to generation, diverting the course of evolution.

This process of replacing damaged genes, or switching them off, is known as germline gene therapy. Although such therapy is technically possible, its use in humans has been constrained by ethical dilemmas and its unpredictable nature. After all, it isn't always possible to know just where the inserted genes will end up, and if they do interrupt other cellular functions, there is a risk of gross abnormalities. However, germ-

line gene therapy in humans has recently come a step closer, with the successful creation of an artificial chromosome. The chromosome was assembled to include the desired genes and then inserted into very young mice embryos. These mice were crossed with normal mice and the next generation was found to have inherited the artificial chromosome. Certainly, inserting a whole chromosome rather than a gene hitched to a virus gives far more control over the outcome. And it could make germline gene therapy in humans much more predictable, and therefore much more tempting.

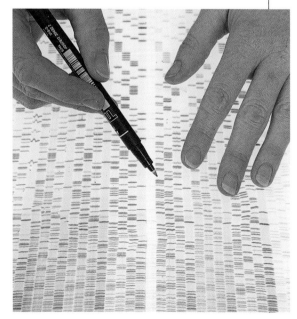

A SCIENTIST *marks reference points on an autoradiogram during DNA sequencing. From this map, the sequence of nucleotides, seen here as dark bands, can be determined in a strand of DNA, in order to find the structure of a set of genes.*

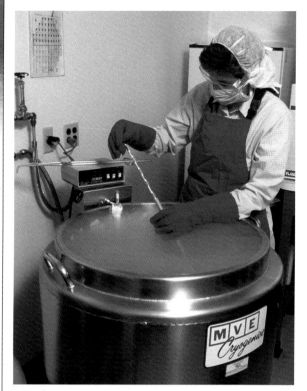

WOMEN CAN NOW BE *induced to ovulate many eggs in one menstrual cycle; the eggs are then removed and put in liquid nitrogen storage like the chamber to the left. This enables women to store young eggs for implantation later in life, reducing the risk of abnormalities.*

Researchers have implanted the nucleus from an older woman's egg into a younger woman's egg, from which the nucleus has been removed. After fertilization, the young cell environment stimulates the nucleus to divide and produce a normal embryo.

Debate continues among the general public, clinicians, and politicians over the far-reaching implications of these technologies. On the one hand, they provide us with an opportunity to have children who are free of any debilitating diseases. On the other hand, these same techniques could be used for non-therapeutic ends, such as choosing eye color, hair color, height, and IQ. While we are far from knowing what, if any, genetic basis there is for such attributes, it is difficult to imagine that people would not be tempted to use this technology once it became more widely available.

IMPROVING FERTILITY

One in eight couples has trouble conceiving or is infertile, and many rely on IVF to have a child. At the present time, an IVF embryo has only a 20 to 30 percent chance of successfully implanting and growing in the womb. This is due in part to genetic abnormalities, although the controversial new genetic screening techniques discussed on previous pages will go a long way to improving IVF success rates.

Techniques that focus on male infertility are also being developed, providing immobile or tailless sperm with the ability to fertilize an egg. In future, infertile men may be able to get their immature pre-sperm cells injected into the testes of other, more fertile males, where the cells would mature and develop normally.

THE FUTURE OF CONTRACEPTION

With world population figures expected to increase by 1.5 billion by 2020, the need for globally acceptable, convenient, cheap, and safe forms of contraception is more pressing than ever. Many existing methods focus far too narrowly on the needs of Western societies. Research is now targeting women in developing countries, who require safe, yet inexpensive, and reversible contraceptives, which do not require a pelvic examination

DELAYING REPRODUCTION

Many women today want to delay having children until later on in life. However, the viability of a woman's eggs diminish with age, reducing her chances of conceiving and increasing certain risks: For example, at 42 a woman is twice as likely to conceive a baby with Down's syndrome as she is at 25. It is now possible to have eggs put "on ice," in facilities similar to sperm storage banks, for implantation at a later time. This shields the eggs from harmful mutations that often occur as a woman grows older.

THE ABORTION PILL

Danco Laboratories was recently given an exclusive license to manufacture, market, and distribute the controversial new drug RU 486 (Mifeprex) in the United States. Mifeprex terminates a pregnancy by blocking the action of progesterone, causing the lining of the uterus to break down so that an embryo will not implant. For women who want another alternative, RU 486 enables termination at a much earlier stage than a surgical abortion; it is also less invasive and causes relatively few side effects. RU 486 inhibits ovulation, too, and thus offers potential as a contraceptive. However, its controversial nature as an abortion pill has impeded research into other potential uses of this compound.

before use, do not affect menstruation, and can also provide protection against sexually transmitted diseases such as AIDS.

The contraceptive pill, based on the progesterone and estrogen hormones, has been used by a large number of women since 1960. Hormone-based contraceptives continue to be developed and are now appearing in patches, gels, and insertable vaginal rings (which are removed from the vagina every three weeks). A progesterone-based Morning Before Pill is also under development, which alters the ion content of the female reproductive tract for 36 hours, making it difficult for sperm to swim.

In the long term, contraceptive drugs will target hormone receptors rather than alter hormone levels. Research into one receptor aims to trick the egg into behaving as if it is already fertilized so that it blocks sperm from penetrating. Other research targets hormones that could prevent a fertilized egg from implanting in the uterus. And though not yet widely accepted, female condoms continue to be refined as they protect against sexually transmitted diseases.

CONTRACEPTIVES FOR MEN

A recent international survey found that two-thirds of men would use a male pill if it were available. One of the barriers to a male contra-

A SINGLE human sperm travels through a fallopian tube of a female in this colored scanning electron micrograph. About 300 million sperm may begin the journey in the vagina, but only a few thousand survive to reach the fallopian tubes.

ceptive is that men are constantly fertile, producing 1,000 sperm a minute. Despite this, male contraception is an active area of research, and a male pill is expected to be available for general use no later than 2010.

The first male pills being tested are based on combinations of the hormones androgen and progestogen. Progestogen works by blocking testosterone production while androgen blocks sperm development, but both hormones have side effects. It is hoped that these hormones, given in the right combination, will prevent fertility while avoiding both loss of libido and potentially harmful changes in fat metabolism.

As opposed to inhibiting sperm production, long-range strategies involve allowing sperm to mature normally, then altering their makeup so they cannot fertilize an egg. New work on contraceptive vaccines could develop a vaccine that interferes with a sperm's functionality in this way. Ultimately, however, any future developments in contraception will depend on the willingness of the pharmaceutical industry to invest in further contraceptive research.

Mending the Broken Brain

Depression will be the second leading cause of disability worldwide by 2020, while Alzheimer's disease will reach near-epidemic proportions. New drug and gene therapies promise to finally subdue mental disorders, however.

Angry gods, bad mothers, the evil eye—at one time or another, contemporary wisdom fingered each as insanity's cause. Even as late as the 1970s, debate raged over whether nature or nurture gave rise to mental ills. But the technological advances of the past three decades have rooted mental illness firmly in the brain.

High-resolution microscopes now reveal the traces of illness in postmortem cells. Imaging studies record aberrant spurts of activity in living brains, while DNA sequencing techniques promise to uncloak the genes that contribute to mental illness. These technologies will become ever more powerful in the 21st century. As researchers use them to mine the human genome and to chart the neural disarray that

MENTAL ILLNESSES *isolate sufferers from themselves and others. New drugs that more precisely target the brain areas devastated in these diseases promise better treatment.*

scars the mind, they will discover new ways to mend the broken brain.

RECEPTIVE RECEPTORS

Treatment advances have already transformed life for many psychiatric patients. The first treatments for Alzheimer's disease emerged in the 1990s, as did a new class of drugs for schizophrenia. These, the atypical antipsychotics, quieted disordered thoughts and hallucinations without causing the serious movement disorders associated with earlier medications. Similarly, Prozac and its sister SSRIs (selective serotonin-reuptake inhibitors) more effectively reawakened hope than earlier antidepressants. But these drugs are not cures. The atypical antipsychotics, for example, don't alleviate the lack of motivation, dulled emotions, and foggy reasoning common among schizophrenics.

Nor are these newer medications without side effects. Although more selective than their

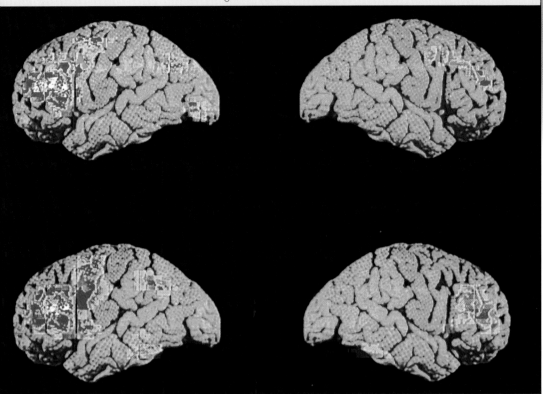

predecessors, they still inundate brain areas unassociated with psychiatric symptoms, causing problems such as weight gain and nausea. The next generation of drugs promises more precise treatment thanks to insights into the great variety of receptors in the brain.

Receptors are a kind of molecular docking platform, the places on a neuron's surface where chemical messengers called neurotransmitters lock on to deliver their instructions. Those instructions tell the receiving cell what to do, giving rise to the moods and musings that mark our days. The same neurotransmitters that modulate joy and sadness also help to regulate digestion, movement, and other physical functions, however. And there's the rub. Most psychiatric drugs work by latching onto receptors so that neurotransmitters can't. In this way, they alter the messages flowing through the brain and, with them, the patient's physical as well as mental life.

Scientists now know, however, that most neurotransmitters have many subtypes of receptors, not just the one or two originally thought. Dopamine, a neurotransmitter implicated in schizophrenia and bipolar (manic-depressive) disorder, has five known receptors, while serotonin, a major player in depression, has 15. GABA (gamma-aminobutyric acid), the brain's main inhibitory neurotransmitter and a suspect in many mental ills, is an exception with two. By selectively targeting just those receptor subtypes impacted by a particular

RESEARCHERS USE POSITRON *emission tomography (PET) to eavesdrop on the living, working brain. The above scans reveal clear differences in the patterns of activation in a normal subject's brain (top) and in a schizophrenic's (below).*

mental illness, researchers hope to develop safer, more effective and faster-acting treatments.

Automated systems that rapidly identify, synthesize, and test potential new drugs are streamlining the process of discovering treatments that are effective. These drug-discovery systems let scientists test as many as 15,000 chemicals per month to see whether they bind to specific receptors. Those that do can then be studied to determine whether they curb the symptoms of mental illness. New drugs discovered this way can point researchers to promising neurotransmitter systems to study. As often, however, the converse is true: Researchers discover new targets for drugs as they plot the paths of mental illness in the brain.

MISSED CONNECTIONS
Most of the prevalent psychiatric disorders leave but a faint footprint upon the brain. Alzheimer's disease is the primary exception. It destroys cells, twisting the fine fibers of neurons into neurofibrillary tangles and heaping debris from attacked cells into brain-clogging deposits called plaques. Scientists can see these changes in post-mortem brain tissue.

Such gross violations don't mar the brains of people who have had depression, schizophrenia,

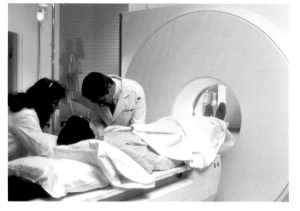

TOMORROW'S IMAGING MACHINES *will both scan the brain and record its electrical firings. They will also allow researchers to measure activity at brain speed, which today's machines (left) cannot do.*

or bipolar disorder, however. These diseases—the three major mental illnesses—cause brain changes as subtle as their symptoms are profound. They may destroy some neurons, but cellular assassination is not their primary mode of operation. Rather, they alter the intricate neural connections that underlie all feeling, thought, and action. The brain cells themselves are normal but their numbers and the connections between them are not.

Take schizophrenia, for example. Patients with this disorder cannot filter incoming information. A deluge of stimuli overwhelms their brains, giving rise to disordered thoughts, blunted emotions, hallucinations, and delusions. In one of many recent insights into the brain abnormalities underlying these symptoms, Francine Benes and colleagues at McLean Hospital, Belmont, Massachusetts, in the United States, discovered too many excitatory neurons and too little inhibitory activity to control them in part of the brain responsible for integrating thought and emotion.

The brain's exquisite system of checks and balances fails in this region, the anterior cingulate cortex, because the cells that would ordinarily inhibit activity are themselves excessively

inhibited: An abundance of inhibitory dopamine fibers connect to the GABA neurons, the brain's main breaks. Benes and others have documented defects in the GABA system in patients with bipolar disorder as well, suggesting that this particular neurotransmitter is a promising target for new treatments.

THE ONE-TWO PUNCH

Discoveries such as the above support the current hypothesis about why the brain's neural wiring short-circuits. Researchers now believe that stress during early brain development, coupled with a genetic predisposition, confers vulnerability to the major mental illnesses. In schizophrenia or bipolar disorder, that stress may be a prenatal virus or obstetrical complications. People with these disorders are more frequently born in late winter or early spring and disproportionate numbers of them were difficult deliveries. Since brain levels of dopamine surge in response to stress, such an assault at a critical point in brain development could cause the dopamine irregularities researchers

NEW CELLS FOR OLD

An eye drop of cells infused into the brain may one day vanquish Alzheimer's disease. Theoretically, the implanted cells would migrate to where they were needed, form connections with existing cells, and begin churning out neurotransmitters, replacing neurons lost to the disease. Researchers have already tried such cellular implants for Parkinson's, another degenerative brain disease. Parkinson's is an easier target because it affects just one brain region, whereas the damage Alzheimer's causes is diffuse. While cellular implants have cured mice with Parkinson's-like syndromes, results in humans have been less encouraging. For some patients the cells have worked too well, causing devastating movement disorders. For others, they have not worked well enough. With fine-tuning, however, cellular implants should one day live up to their promise as a major treatment of the 21st century.

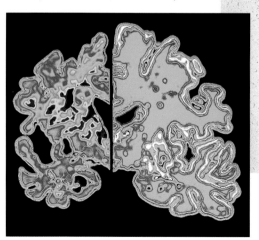

THE BRAIN-SHRINKING *impact of Alzheimer's disease is painfully apparent in these computer images of the same area from an Alzheimer's brain (left) and a normal brain (right).*

see, creating a hypersensitive brain. When dopamine neurons become still more abundant because of normal changes in brain development during adolescence and early adulthood, that sensitivity is further heightened. A stressful event such as moving away for college or starting a new job then triggers the disease.

Researcher Charles Nemeroff and colleagues at Emory University, Atlanta, in the United States, propose a similar pattern for major depression. They believe that early childhood trauma such as neglect or abuse pushes neurons that secrete corticotropin-releasing factor (CRF) into permanent overdrive. This in turn stimulates overproduction of the body's major stress hormone. Because their brains are on perpetual high alert, individuals with a genetic predisposition to depression may spiral into despair in response to even a minor stress.

THE GENE HUNT

Of course, not every abused child becomes depressed and not every severely depressed adult suffered childhood trauma. Different levels of genetic vulnerability confer different degrees of protection or risk. The same is likely true for schizophrenia and bipolar disorder. Nabbing the genes that contribute to these disorders thus remains one of the biggest challenges researchers face.

Despite numerous reported sightings, those genes remain elusive. This isn't surprising. Several genes in combination most likely produce a vulnerability to mental illness. These genes may be common ones that are harmless alone but devastating in combination. Or, they may remain dormant unless other genes or environmental conditions trigger them. Traditional studies, in which geneticists study families with many affected members, provide too few subjects to detect such subtleties. Thus, geneticists are turning to association studies, in which they analyze genetic samples from large numbers of unrelated people who have the disease. The map of the human genome, biochips that enable researchers to detect variations in single genes, and rapid DNA analysis techniques are making such studies possible.

Once known, the culprit genes and their products should lead to not only new treatments but also the first biologically based diagnostic tests for mental illness. Such tests will enable early screening and intervention to protect the vulnerable, developing brain and prevent the anguish and isolation of mental illness.

GENE THERAPY *may one day correct the defects that predispose people to mental illness. Researchers have already tried using it to combat the memory loss of Alzheimer's disease. The scanning electron micrograph below shows a single gene.*

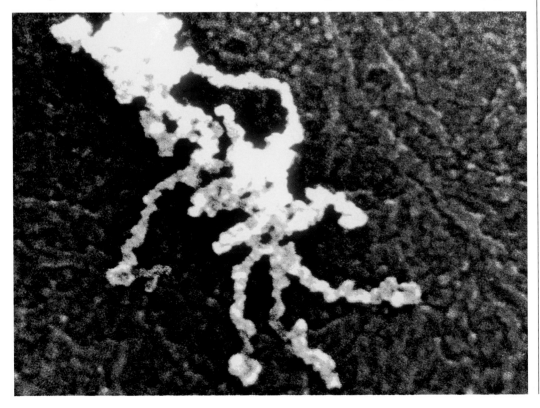

Memory

New drugs to counter the memory loss of Alzheimer's disease will undoubtedly find their way into the neurons of the young and healthy. But the question is, should they?

To control memory is to alter not only what we know but also who we are. Our sense of self rests on the scaffold of our remembered past. One set of memories confirms our claim to kindness; another, our conviction that we are funny or brave. The very selectiveness of these recollections shapes the contours of identity.

Within the coming decades, a new class of drugs called cognitive enhancers will give us the power to selectively fix the rush of experience, thereby altering memory and, with it, ourselves. These drugs are being designed to counter the devastating memory loss of Alzheimer's disease and the milder, but still disrup-

THE HIPPOCAMPUS *(purple) is where the brain transforms experience into memory. Emotional experiences release a flood of hormones that supercharge the hippocampus and sear the brain. Researchers hope to develop drugs that do the same.*

tive, decline that can accompany aging. They will work by altering the molecules that brand experience in the brain.

THE MOLECULES OF MEMORY

Researchers have long known that memory is inscribed at the synapse—the minute gap between cells at which neuronal conversations take place. The more often we gaze at a loved one's face, drive to work, or recite "Ode to a Nightingale," the more often a specific set of neurons fires, strengthening the synaptic connections between cells. These well-worn neuronal pathways are the cart tracks of memory.

Researchers are now trying to decipher the molecular changes that lay down those tracks. They are primarily studying the hippocampus, the seahorse-shaped cluster of brain cells that converts what we learn into what we remember. Of the numerous molecules likely to be

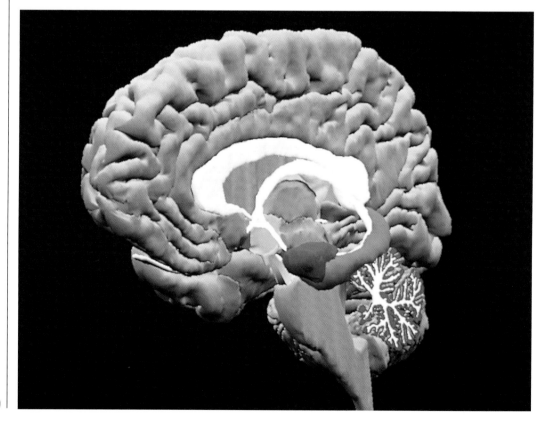

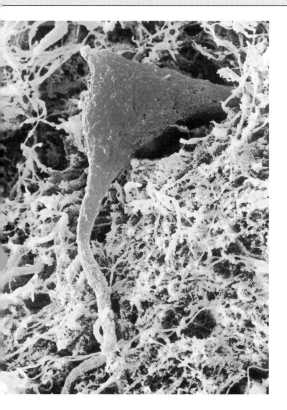

A NEW BREED OF DRUGS *called cognitive enhancers may one day give gamblers (above) and others near-perfect recall. They will work by strengthening communication between neurons (left, red) in the brain regions that lay down memory.*

involved in this process, they have identified at least two that may provide promising targets for drugs.

The first is a receptor called NMDA, which perches on the membrane of cells at the receiving end of neuronal conversations. When it is activated, the NMDA receptor opens. This initiates a chemical process called long-term potentiation (LTP), which strengthens the synapse. LTP can last for days, or even weeks, and is believed to be central to encoding many types of memories. Mice that have been genetically engineered to have souped-up NMDA receptors cream their peers on memory challenges. Conversely, those with compromised receptors soon forget their lessons. Drugs called ampakines that boost LTP are among the cognitive enhancers now in development.

The other target for memory-boosting drugs is CREB-1, a molecular master switch that turns on numerous genes, some of which produce proteins involved in cementing memory. These proteins alter the structure and activities of neurons to create new synapses or strengthen existing ones. CREB-1 interacts with a sister molecule, CREB-2, that inhibits memory consolidation. The relative strength of these two molecules appears to influence how easily long-term memories form. Where a pill that boosts CREB-1 activity could enhance memory, one that increases CREB-2 could induce forgetfulness, potentially sparing a trauma victim the horror of reliving the event time and again.

EVERYONE A MR. MEMORY?

These newly identified molecules of memory are but the first potential smart drug targets. Researchers are just beginning to understand the complex processes through which our minds form, consolidate, store, and retrieve memories and to map the intricate circuits involved. The hippocampus is but one brain region critical to memory, for example. The cortex, in which memories are stored, and the amygdala—part of the brain's emotional circuitry now known to modulate the strength of memory—are others. As researchers delineate the interactions between these regions, they may discover additional opportunities to enhance the power of recall.

Just how much we may want to remember is an open question, however. Jumpstarting a healthy mind may not be as desirable as it sounds. Most of what we experience and learn lands in the recycle bin for good reason. The mnemonic feats of Mr. Memory have little value in our day-to-day world, and life is kinder when recollections of grief, humiliation, and regret remain muted. Since we can't control what wafts into consciousness, we might rue what we experience while we are memory-enhanced. Memory boosters also raise obvious questions of equity, not just among students cramming for exams, but among candidates debating the issues of the day, chess masters contemplating the next move, or even gamblers calculating the odds.

Such concerns won't halt the search for smart drugs. Nor should they. Cognitive enhancers are critical to slowing Alzheimer's disease. Whether they are used for other purposes will be for society to decide.

41

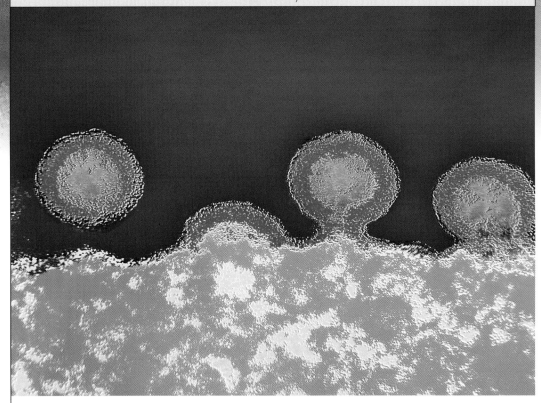

Fighting Disease

Our most potent protection against disease comes not from

the pharmacist's shelf but from our own immune system,

which physicians of the 21st century will harness to fight

illnesses from cancer to AIDS.

The war within is ceaseless. The ever-vigilant cells of the immune system constantly scour the body, devouring viruses, annihilating bacteria, and pushing mutated cells to suicide. Despite these prodigious defenses, however, disease too often gets the better of us. HIV/AIDS hijacks and destroys the very cells that sound the immune system's alarm. Cancers arise from within, causing neither the tissue damage nor the inflammation that ordinarily stimulate a vigorous immune response. Some infectious diseases, such as Ebola, invade too rapidly for our bodies to mount a successful defense.

Yet hope is growing that a new generation of vaccines can bolster the body's response to these and other ills. The new approaches draw on a deeper understanding of the immune system, advances in genetic engineering, and new methods of drug delivery to either inoculate against or provide treatment for conditions as varied as cancer, cocaine addiction, multiple sclerosis, Alzheimer's disease, and AIDS. Different though their targets and the methods of creating them are, these new vaccines ultimately do what vaccines have always done: they teach the immune system to recognize and destroy the body's attackers.

APT PUPILS

Vaccines capitalize on the immune system's capacity to learn. If an invader, or antigen, slithers past the stronghold of the skin or through the sticky lining of the respiratory and digestive tracts, it encounters hordes of phagocytes—white blood cells that engulf any and all foreigners without discrimination. Some of these phagocytes dice up their victims and present pieces of them to the immune system's sharpshooters—powerful B cells and killer T cells, which respond only to specific antigens. When they meet their match, B cells produce

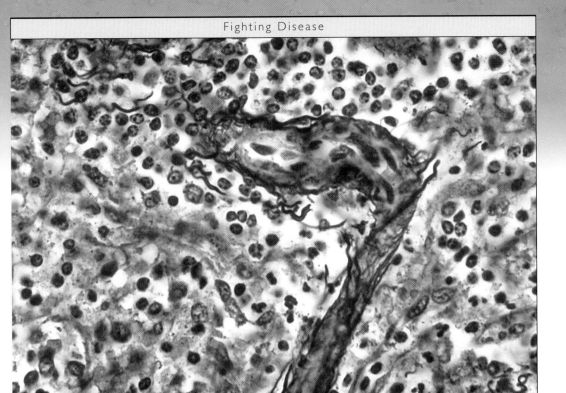

antibodies that stick to antigens as they travel through the body's fluid-filled spaces, marking them for destruction. Killer Ts destroy infected or mutated cells directly, ridding the body of pathogens that have found their way into cells, as well as of potentially cancerous cells. B and killer T cells never forget an enemy and thus are the basis of lifelong immunity. After a battle, they leave a platoon of antibodies and special memory cells behind, ready to attack quickly should a specific antigen show its face again.

Traditional vaccines use dead or weakened forms of a virus to stimulate this cascade of events without causing infection. But the traditional approach doesn't always work. Consider the problem of developing a vaccine against HIV, the virus that causes AIDS. The virus is too lethal to use in a weakened form, given the slim possibility that it could mutate back to virulence. Killed viruses, however, can't infiltrate cells. They activate the humoral arm of the immune system, the B cells, but not the T cells that are essential to fighting AIDS. To complicate matters further, HIV, like influenza, has many subtypes. A given individual may harbor several forms of the virus in different parts of the body. How can a single vaccine protect against them all?

GENE VACCINES
Genetic engineering may provide the solution to developing a successful AIDS vaccine and

A COLORED TRANSMISSION electron micrograph (above, left) shows the HIV virus budding from an infected helper T-lymphocyte (at the bottom of the micrograph in pink), the type of cell that ordinarily sounds the immune system's alarm. Four viruses are shown at different stages of budding. At center left, the virus acquires its coat from the cell membrane (red); at right, the virus buds from the cell; at center right, budding is almost complete; at left, the new virus is free floating. Once it is free, the virus infects other helper T cells. Helper T cells are the part of the body's immune system destroyed by HIV, the virus that causes AIDS. Spleen tissue (above, magnified 160 times) is a reservoir of immune cells, including T cells, B cells, and phagocytic cells, which engulf and digest not only antigens but also worn out red blood cells.

more. Scientists can now take a pathogen and isolate genes that excite the immune system but do not cause disease. They then insert these genes into harmless loops of bacterial DNA called plasmids. When injected into the body, the plasmids enter cells and the genes they carry then go to work, churning out the pathogen's proteins. These alien proteins then rouse the immune system, which attacks the plasmid-infected cells as well as any others displaying the malevolent proteins.

In this way, DNA vaccines could treat as well as prevent disease. Since plasmids can carry genes from tumors, cocaine, plaques that form during Alzheimer's disease, or just about anything, they could sensitize the immune system to these substances, triggering a response against existing disease.

43

Researchers expect to complete the first wave of clinical trials for DNA vaccines by 2010. They are targeting infectious diseases, including hepatitis B, herpes simplex, flu, malaria, tuberculosis, and HIV, as well as developing therapeutic vaccines for HIV, Alzheimer's disease, cocaine addiction, and various forms of cancer. While many experimental cancer vaccines use the approach described above—splicing a specific bit of protein that occurs only on tumor cells into a plasmid—others tinker with entire tumor cells. Researchers harvest the patients' own cancer cells, render them harmless through irradiation, insert a molecule known to excite an immune response into the cells, then return the altered cells to the patients' body. The first such vaccine, AVAX Technologies, Inc.'s M-Vax, is now available in Australia to treat late-stage melanoma, the deadliest form of skin cancer, the incidence of which is increasing rapidly.

BOOSTER SHOTS

Promising though DNA vaccines are in theory, they have shown mixed results in practice. Too often, they don't elicit an adequate immune response. The possible reasons are many. The genes used, selected from among the thousands in a pathogen's genome, may not be virulent enough. The dosage or delivery schedule may not be optimal. The killer T cells, although activated, may not attack. The vaccine may not work for long enough—most stop producing significant amounts of protein within a month.

As scientists learn more about the workings of the immune system, however, they are finding ways to boost the potency of DNA vaccines. They have learned, for instance, that they can manipulate the plasmid itself to provoke a stronger immune response. They can also add a "kicker" to the genetic brew. Several substances are now known to enhance the immune system's ability to destroy cancer cells, for example. By hooking one of these to a tumor antigen, researchers hope to shake the immune system into action. Yet another approach delivers a one-two punch by administering a DNA vaccine to stimulate T-cell activity, followed by a conventional one to elicit antibodies. This "prime-boost" strategy is now being tried against HIV.

DNA vaccines aren't the only new kids on the block, though. Equally interesting are vaccines based on components of the immune

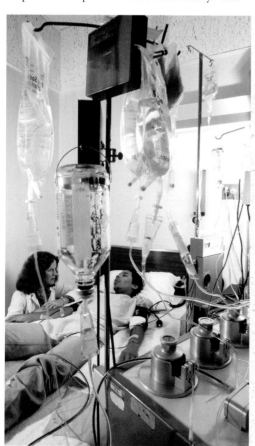

MISTAKEN IDENTITY

In biology as in politics, the consequences of confusing friend for foe can be disastrous. When the immune system makes such a mistake and attacks the body's own tissues, diseases such as type 1 diabetes, multiple sclerosis, and rheumatoid arthritis result. Current treatments, such as blood transfusions for multiple sclerosis (left), have limited success, and researchers now hope to stop the progression of such diseases by retooling the immune system. One approach being developed to treat multiple sclerosis vaccinates patients with irradiated versions of their own disease-causing cells. The irradiated cells are perceived as foreign, triggering an immune response that theoretically will wipe out all similar cells in the body. Another treatment capitalizes on the paradoxical observation that ingesting self-antigens (body proteins that are mistaken for foreigners) can sometimes suppress the immune response to them. Scientists are developing edible diabetes vaccines based on this principle. While it's too soon to predict whether these specific efforts will result in effective treatments for autoimmune diseases, the strategy of retraining the immune system one day will.

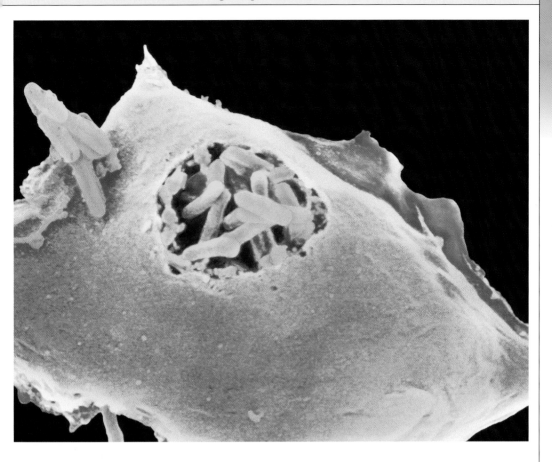

A MACROPHAGE CELL *(yellow) can be seen in this colored scanning electron micrograph. The macrophage cell is engulfing the bacteria* Mycobacterium tuberculosis *(green), a process known as phagocytosis.* Mycobacterium tuberculosis *bacteria are spread by coughing or sneezing. Once inhaled into the lungs, they form an infected lesion, causing pulmonary tuberculosis (TB, or consumption), which can be fatal. The macrophage cell is a scavenger immune cell that migrates to the site of infection in the body. It protects by removing bacteria or other foreign bodies from blood or tissues.*

system. Some of these involve infusions of antibodies or cytokines, chemicals that help modulate immune functions. Others use the body's dendritic cells. These specialized cells are the immune system's sentries. Stationed throughout the body, they swoop down on antigens, dismember them, then dash to the lymph nodes, where they display their booty, thus activating the appropriate killer T cells. Researchers can now harvest a patient's dendritic cells, grow them in culture, expose them to an antigen, and then release this army of primed cells in the body. As with DNA vaccines, trials of dendritic cell vaccines have had mixed results, but the field is still very young.

SPECIAL DELIVERY

New methods of delivering vaccines also promise to improve health this century. Time-release vaccines could eliminate the booster shots now needed to provide sufficient immunity against diseases such as tetanus or diphtheria. In one approach to this problem, scientists at Osiris Therapeutics, Baltimore, Maryland, in the United States, are transforming immature cells in the body's bone marrow into vaccine machines. By linking vaccines to genes turned on at different points in the cells' lifecycle, they can control when the vaccine is produced, and thus provide immunity over a longer time.

Other researchers are devising ways to stop pathogens where they attack. For example, vaccines that target the mucosa—the moist lining of the respiratory, gastrointestinal, or genital/urinary tracts—may provide more potent defenses against pathogens that enter the body through these tissues. Among the vaccines under development are a nasal spray for influenza and genetically engineered potatoes that protect against *E. coli* bacteria.

Edible vaccines, such as these potatoes and goats' milk genetically modified to prevent malaria, hold enormous promise for improving global health. They could provide a low-cost, easy-to-administer method of inoculating large numbers of people, especially in poorer countries. The admonishment to "eat your vegetables" might soon take on a whole new meaning.

Cures and Reversals

A diagnosis of cancer or heart disease, multiple sclerosis or stroke may warrant little more than a shrug once researchers learn how to manipulate the body's own repair mechanisms.

*I*ncurable is a terrifying word. It summons fear of debilitation, of senses permanently silenced, and of a life ended all too soon. But the disorders that terrify one generation often cause little concern for the next. In most of the world today, the thought of polio no longer quickens the heart—nor smallpox, nor diphtheria, nor plague.

Just as we take these illnesses in our stride today, so our children's children—or even our children themselves—will look upon a number of today's dreaded diagnoses with nonchalance. And while gene therapy may one day conquer cancer, and a deeper understanding of the immune system may obliterate AIDS, there are other advances even nearer at hand. If today's most promising research pans out, cancer will be transformed into a chronic but manageable disease. Most strokes will cause little permanent harm. Diabetics will churn out their own insulin, while arthritics swing painlessly around ballroom floors. Even blindness and paralysis will no longer confer life sentences.

THE BODY'S REPAIR KIT

While the last century's greatest medical breakthroughs came from vanquishing infection, a great many of this century's will come from manipulating the body's own repair systems, starting with stem cells.

Stem cells—the immature cells from which all others arise—form in the earliest stages of embryonic development. As development proceeds, embryonic stem cells begin to specialize. Some may form the skin stem cells; others may become blood stem cells, neural stem cells, and so on. These specialists then go on to spawn all the tissues of the body.

As adults, we retain reservoirs of specialized stem cells that contribute to the body's maintenance and repair. Some adult stem cells routinely replace worn-out blood cells or give rise to the osteoblasts that build and mend our bones. Others replace damaged muscle, brain, or liver cells. But they are neither numerous nor active enough to overcome the ravages of severe trauma or illness. What's more, their numbers may diminish as we age.

Stem-cell therapy aims to increase the healing power of our native stem cells by augmenting their numbers. Harvested from an embryo or adult, grown in the lab, then transplanted back into the body, these stem cells can potentially reverse paralysis, diabetes, blindness, multiple sclerosis, Parkinson's disease, and more. They are doing so now, in rats and mice.

Stem cells injected into the brains of rats that had suffered severe strokes improved their mobility. Stem cells transplanted into the eyes of rats with damaged retinas began developing like healthy retinal cells, although scientists could not say whether these cells would ultimately function in the way they should. And embryonic stem cells coaxed into becoming myelin–producing brain cells in the lab produced normal-looking myelin

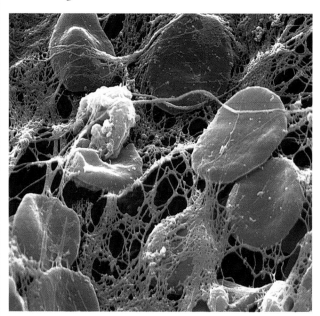

BLOOD CLOTS *form to repair damaged blood vessels (above). Angiogenesis-based drugs will promote healing by stimulating the growth of new blood vessels which will rush oxygen, nutrients, and white blood cells to injured tissues.*

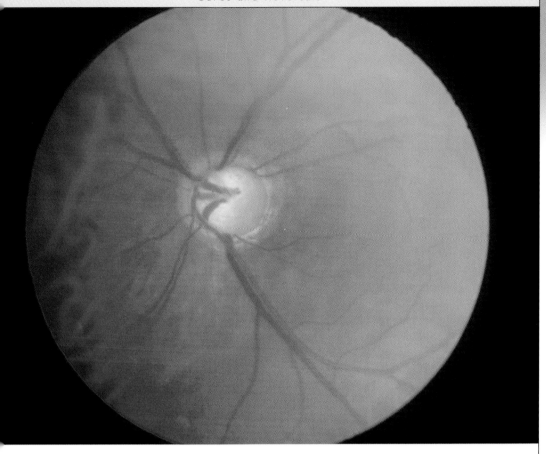

THE BODY CONTAINS AN ABUNDANCE *of growth factors and cells that aid in healing. Researchers hope to manipulate these to treat conditions as disparate as paralysis and blindness. Potential treatments include stem cells to reverse nerve damage that can lead to blindness (above) and growth factors to jumpstart myelinated nerves (right) that are often left intact but nonfunctioning after a spinal cord injury.*

when transplanted into mice, suggesting a possible treatment for multiple sclerosis.

UNANSWERED QUESTIONS

Promising though these results are, scientists still don't know whether what works in rats will ultimately work in people or whether the implanted cells will continue to function. Nor do they understand the basic biology underlying stem cells' actions. One example: Stem cells in the animal studies behave like biological emergency medical technicians, rushing off to where they are needed. In the eye study, for example, the stem cells developed into retinal cells in damaged eyes but not in healthy ones. Similarly, stem cells injected far from the site of injury in brain-damaged rats migrated to the damaged area, improving the animals' impaired coordination. Although researchers can implant stem cells without understanding what controls their migration, better understanding will yield far more effective treatments.

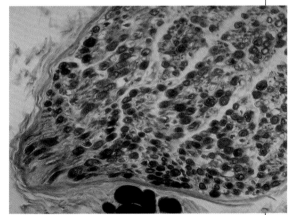

The best source of stem cells is an issue, too. Embryonic stem cells are the most prolific and versatile. They replicate indefinitely and can theoretically be coaxed into forming any cell in the body. But controlling the cells' growth so that they don't form tumors or spontaneously mutate—laying down skin cells where heart muscle was wanted, for example—is a challenge, as is countering rejection. The ethical objections to extracting cells from embryos may present the greatest barrier of all.

Adult stem cells could be harvested from the patient's own body. This would eliminate rejection, but adult stem cells are generally less malleable and shorter-lived than embryonic ones and are often difficult to isolate and purify.

Another problem is that scientists have yet to locate the adult stem cells for some tissues, including such critical ones as the heart.

Despite these challenges, researchers anticipate initiating clinical trials of stem-cell therapies within the first decade of the 21st century. And they're making good progress in overcoming obstacles. Researchers at the University of Cambridge, England, have successfully prolonged the life of adult stem cells in culture, while colleagues at the Salk Institute, La Jolla, California, reported a new source of stem cells—fresh cadavers. Even more intriguing, adult stem cells may not be as set in their ways as first thought. Scientists have transformed easy-to-access bone marrow stem cells into brain, liver, and muscle. Such cellular alchemy could provide a ready supply of materials with which to cure currently incurable conditions.

STARVING TUMORS

Like stem cells, angiogenesis, the growth of new blood vessels, is one of the body's repair mechanisms and a promising source of future treatments. Indeed, the Angiogenesis Foundation in Cambridge, Massachusetts, predicts that angiogenesis-based treatments will be to the 21st century what antibiotics were to the last. Just as antibiotics attack a common feature of disparate infectious diseases, so treatments that manipulate angiogenesis target a contributor to many of today's most lethal and

CANCER CELL GROWTH

Cancer tumors need blood to thrive. They get it by establishing their own blood supply, as shown below (the cancer cells are green). Drugs now in clinical trials impede this process and promise to transform cancer into a manageable disease. Stem cells may also play a role in future cancer treatment.

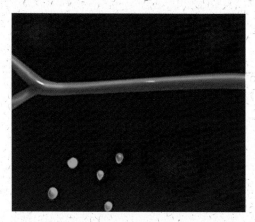

1. Implanted cancer cells glow with green fluorescence protein.

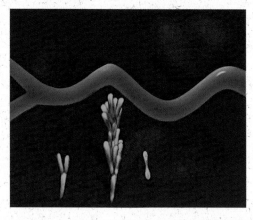

2. Three of the original cancer cells have survived to begin replicating. Signals cause them to grow toward the vessel.

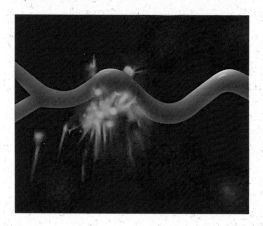

3. The cancer cells reach the existing blood vessel.

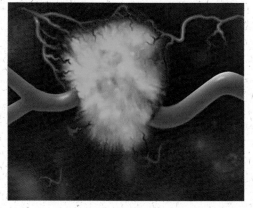

4. When they number only 100–300, the cancer cells have created new, fully functioning blood vessels.

disabling disorders—the growth, or lack thereof, of new blood vessels.

In the healthy body, angiogenesis takes place only to promote healing and to support fertility and pregnancy in women. But in conditions ranging from diabetic blindness to rheumatoid arthritis to cancer, the balance of growth and inhibitory factors that normally keeps angiogenesis in check fails. New blood vessels sprout, extinguishing vision, inflaming joints, feeding hungry tumors, and supplying escape routes for cancer cells. By developing drugs that inhibit angiogenesis, scientists hope to halt the progression of these diseases, at the very least stabilizing, if not in fact curing, patients.

Stopping blood vessel growth is only half the angiogenesis story, however. Damaged tissues need increased blood flow to heal effectively. Without it, wounds fester or tissue traumatized by a heart attack or stroke dies. Drugs that encourage angiogenesis could revive dying brain cells in the hours following a stroke. And they could create a coronary artery bypass without surgery or supply the rush of blood needed to stimulate the healing of a slow-to-heal wound, such as a diabetic foot ulcer.

Researchers have now identified more than 19 angiogenesis growth factors and 30 inhibitors in the body. The first commercial, approved product based on these chemicals, a gel that promotes wound healing, came to the market in the late 1990s. By the year 2000, some 40 anti-angiogenesis treatments for cancer were in clinical trials. Preliminary results suggest that some of the new medicines could effectively shut down a tumor's blood supply, thereby halting its growth and, in some cases, even shrinking it. While such treatments are not cures, at the very least they promise to transform cancer into a far more manageable condition and to do so without major side effects.

Between them, angiogenesis-based treatments and stem-cell therapy make possible a future in which paralysis, heart disease, blindness, diabetes, cancer, stroke, and many other conditions are significantly less threatening than they are today. For people like Christopher Reeve, the American actor paralyzed from the neck down in a riding accident, we can't get

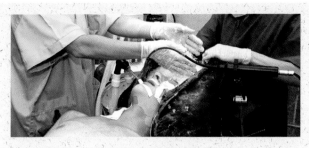

COOL HEADS

The advice to "chill out" is taking on a whole new meaning for victims of ischemic strokes, caused when a blocked artery cuts off the blood flow to part of the brain. Lowering the brain's temperature by as little as 1.8° F (1° C) in the hours immediately following a stroke reduces brain damage and deaths. Norwegian researchers enveloped 17 patients in cooling blankets right after they'd had a stroke. Six months later, these patients had twice the survival rate of a control group whose members had received only conventional treatment. Cooling the entire body carries risks of its own, however, and researchers are now developing devices designed to cool only the brain. One approach slips a cooling pad between the skull and the brain; another chills the blood feeding the brain (and thus, eventually, the damaged brain itself). This cool therapy promises to be but one in a battery of new treatments that vastly reduce the disability now associated with strokes.

there soon enough. Reeve once quipped that he wished he were a rat so that he could get the latest treatment for paralysis. Although the wait is frustrating, the long journey from animal research to human treatment is necessary to assure that physicians uphold their fundamental duty to first do no harm.

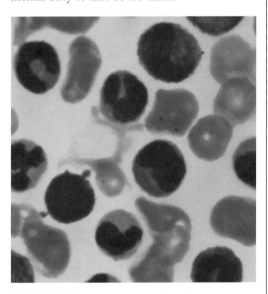

NEW EVIDENCE *suggests that even "liquid" cancers such as leukemia (shown invading bone marrow, above) use angiogenesis to grow. This means that anti-angiogenesis drugs might work against them, as well as against solid tumors.*

The Medicine of Body Image

The ever-growing obsession with body image

demonstrates a fixation with youthful perfection.

To what lengths will we go in order to obtain this ideal?

There is an abundance of products and treatments, surgery and drugs, available today, all promising slimmer stomachs, bigger muscles, straighter teeth, younger hair, quicker minds, and, of course, perfect breasts. Already, we can look younger at 50 than our mothers or fathers did at 35. And the pressures to do so are immense.

An American woman is surrounded by air-brushed images of ever-younger models who, on average, weigh 23 percent less than she does; 25 years ago, models weighed just eight percent less than the national average. Men, too, are constantly bombarded with images of the air-brushed and steroid-enhanced bodies of models and actors. The result is that today, up to 35 percent of all Americans are trying to lose weight, and there are epidemic levels of the slimming diseases bulimia and anorexia among teenagers. Cosmetic surgery is also attracting an ever-younger clientele—though not without

some controversy, as occurred when a 15-year-old girl in Great Britain was promised breast implants for her 16th birthday by her parents, both of whom were plastic surgery advisers. In addition, 10 percent of all procedures in the United States are now performed on men.

That is not to say that this obsession with youthful perfection is without reason. A *Wall Street Journal* study indicated that younger and more attractive job applicants were more likely to be hired over other applicants with the exact same credentials. And our desire to change the way we look is greatly encouraged by a vast industry—the diet industry is worth almost $50 billion annually in the United States alone.

COSMETIC SURGERY

New, non-surgical cosmetic procedures mean that frown lines can be injected with collagen or botulism toxin, and aged skin "resurfaced" with a carbon dioxide laser, in the same time it takes to have a haircut. Similarly, cosmetic surgery is becoming far cheaper and safer; the healing process is quicker and the results appear more natural, as procedures become ever more

WHEN THE PRESSURE IS ON *to go shopping for the perfect body, perfect teeth, and perfect mind, will we all give in, and if we do, will it then be possible to tell one another apart?*

BREAST AUGMENTATION *is booming among teenage girls, according to many plastic surgeons. Today, saline-filled breast implants (below) are used, but in the future less-invasive surgery may involve breasts grown in the lab from a patient's own fat cells and then infused back through keyhole slits in the chest. Non-surgical cosmetic procedures are becoming more common, and it's now possible to get frown lines injected with collagen (right) in less than an hour.*

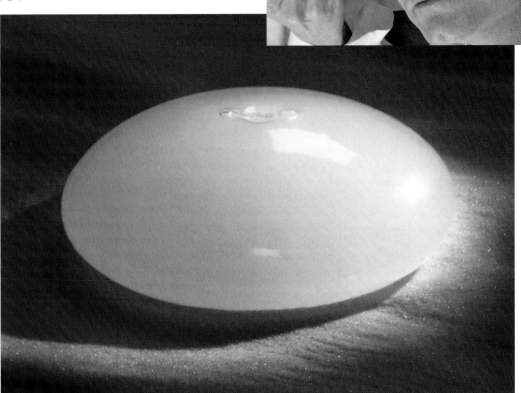

sophisticated. For example, endoscopic lifting is a new surgical face-lift technique where half-inch (12 mm) slits, rather than major surgical incisions, are made around the hairline. Video fiber optics are then used to lift, tighten, and reposition the underlying tissue and muscle as well as the skin, so avoiding the stretched look of earlier face-lifts.

There is now a growing demand for more "intimate" cosmetic surgery among younger women wishing to have their genitals modified to appear smaller and less prominent. And the desire to grow taller can be made a reality through a surgical procedure—originally developed for people with limb deformities and dwarfism—which involves slicing through the leg bone then pinning it with a metal frame. Each day, the patient turns screws on the frame to pull the bone apart fractionally. The ends of the bone continue growing to maintain contact with each other and in this way, the leg can be lengthened by as much as 12 inches (30 cm) in just a few months. Although it is still illegal to perform the operation for cosmetic purposes in the United States, it is legal in Russia and more Americans are traveling there to undergo this leg-lengthening procedure.

HAIR REPLACEMENT

The possibility of growing a new head of hair is still a long way off. However, a new technique enables grafts of as few as one to two follicles to be transplanted to bald areas. The new growth looks more natural than previous hair transplant techniques, although it can take up to two years to restore a full head of hair.

Another problem with hair transplants has been obtaining enough hair to transplant, an obstacle which could be overcome if people were able to donate hair. The first successful transplant of donated hair has already taken place between a couple from Scotland. Hair cells taken from the husband's scalp and implanted into a bare patch on the wife's forearm took hold and grew, making donation of such tissue likely in the future.

NEW TEETH

Procedures are being developed to perfect the smile. Having pinpointed several genes that affect normal tooth development, researchers now aim to switch these genes on to grow new teeth, either in cell culture or directly in the gum. The various components of genetically engineered mouse teeth—cementum, enamel, and dentin—have already been grown in the laboratory. Other research is trying to use genetic techniques to regenerate the periodontal ligament, which holds a tooth in place on the bone and which is often destroyed by gum disease.

SMARTENING-UP

The idea of a pill to boost intelligence and memory has existed since the 1970s. Such a drug has great allure, and there are at least 140 chemicals and food additives available which claim to boost exam results and improve work success. While it is true that many such drugs improve alertness and mood in the short term, their long-term benefits for memory and for learning are not certain.

In years to come, enhanced memory and learning may be genetically engineered instead. American researchers have created a "smart mouse" by adding extra copies of the naturally occurring NR2B gene to mouse embryos. This gene fundamentally controls the brain's ability to associate one event with another, a core requisite for learning. The new mouse strain—called "Doogie"—was better able to solve maze tasks, learn from experience, and retain that knowledge. The same gene exists in humans, and while testing on humans is a decade or more away, researchers say this discovery may eventually lead to a therapy for sufferers of Alzheimer's disease. More

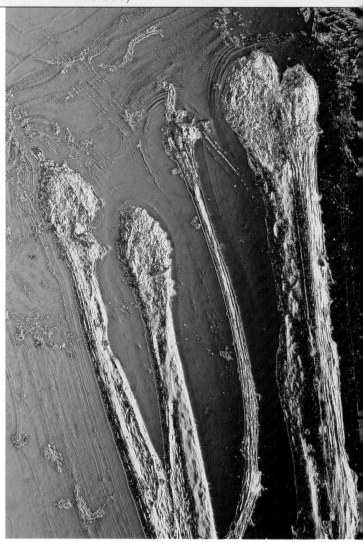

IN THE FUTURE *we may be able to regenerate our own lost teeth and hair. A new hair transplant technique involving grafting one or two follicles (above) to be transplanted to bald areas promises to give more natural looking growth than previous transplants. It may also be possible to replace lost teeth using new gene-based tooth regeneration techniques. These teeth may be grown in the laboratory or even regenerated by the patients themselves using their own tissues. These tissues, researchers say, should possess a "memory" of how the teeth first developed. The components of genetically engineered mouse teeth—dentin, cementum, and enamel (left)—have already been grown in culture in the laboratory.*

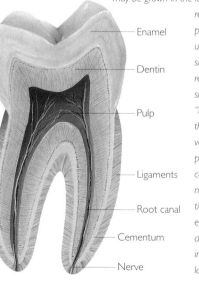

- Enamel
- Dentin
- Pulp
- Ligaments
- Root canal
- Cementum
- Nerve

controversial is the possibility of its application to the wider population, in that it may be possible to manipulate human embryos to contain extra copies of the gene and thus create a smarter race of humans.

Obesity

People in Western societies have become significantly taller and heavier in the past century, largely due to better nutrition. Now, however, over half of all Americans are reported to weigh more than their ideal weight. Sedentary jobs and lifestyles are partly to blame, and with the vast range of foodstuffs available (over 50,000 today, compared with just 500 a century ago), it is becoming increasingly difficult to avoid over-eating. Recent research also suggests that our bodies have a stable weight, which is not determined by calorie intake and exercise but by changes in the nervous system, altering both our appetite and metabolism.

And so the search for a pill to make us thin continues. Some drugs in development aim to control the neuronal signals that tell us we are hungry. Others would allow us to eat as much as we like without putting on weight.

One research focus is on the protein perilipin, which stops fat stored in the body from breaking down. When researchers switched off perilipin in genetically engineered mice, they were found to be thinner and to have higher metabolic rates than normal mice. A different approach being explored is reducing the amount of sugar flowing into cells from the blood, for which the PTP-1B enzyme may hold the key. When PTP-1B is switched off in mice, they gain only half as much weight as control groups of mice fed on the same high-fat diet.

All this research is still to be applied to humans, however, so the ideal obesity drug remains a thing of the future. Indeed, recent experience with obesity drugs suggests that they can sometimes do more harm than good. However, the potential market for a drug that works is vast, and every major pharmaceutical company is now directing at least some research effort into this area.

Treating the Very Obese

For people whose obesity may be life-threatening, a new surgical procedure called laproscopic Roux-en-Y may hold the answer. Staples close off the stomach, except for one small pouch which can hold about two tablespoons of food, and the first and second parts of the small intestine are bypassed, a procedure that dramatically reduces absorption. Most patients lose around 70 to 80 percent of their body weight permanently, although they can suffer side effects such as nausea and do require substantial ongoing supplements of vitamins.

A new, non-surgical procedure for chronic obesity involves inserting a silicon balloon into the stomach and then inflating it. This is intended to create a sense of fullness and thus curb the appetite, although there have been reported problems with some balloons bursting.

The tensions between the fashionable images of thinness and our increasing propensity for weight gain will not be resolved easily. But if society pursues its current, relentless ideal of youthful bodily perfection through the use of drugs and surgery, a substantial number of people of normal size and average appearance may be made to feel abnormal—unless, of course, we all take advantage of these body-overhaul strategies. And if we were to do that, would we really want a world of goddesses and gods, all beautiful, ageless, and identical?

THE ANTI-CHOCOLATE PATCH

A new patch has been developed in Great Britain which claims to reduce the craving for chocolate and other sweets. The vanilla-smelling patch, worn like a nicotine patch, has been effective in dietary trials, halving the subjects' intake of chocolate and reducing their consumption of sugar-based drinks. Manufacturers of the patch are now developing other patches to help reduce cravings for fatty foods and to alleviate premenstrual tension and insomnia.

Methuselah's Secrets

Immortality won't be ours, but longer, healthier lives
are likely in the 21st century as scientists discover
the fountains of youth within our own cells.

W e want, most of all, more time. We crave another year, another day—another hour in which to laugh with a friend, kiss a loved one, or watch a grandchild grow tall. Throughout history, we have sought the magic elixirs that would grant this wish for a long, healthy life. As this century unfolds, scientists are likely to find them in the genes and cellular mechanisms that govern our aging. And their discoveries promise to deliver the gift of longer life to ever more of us.

LIVING LONGER

Just how long our longer lives are likely to be is a matter of dispute. Historically speaking, aging is an oddity. A child born in the United States as recently as 1900 could expect to live

TODAY'S JAPANESE INFANTS *have a life expectancy of 80 years—the longest in the world. During the 21st century, as knowledge of our genetic machinery increases and we make new biological discoveries, ever more of us should have a better chance of reaching a ripe old age, and reaching it healthily.*

only 47 years. But during the 20th century, life expectancy soared in economically developed nations. Vast improvements in sanitation and nutrition, widespread vaccination and use of antibiotics, and better treatments for heart disease, stroke, and other late-life ills added 30 years to expected life spans. A baby born in the United States today is likely to reach age 76; a Japanese infant, age 80, the longest life expectancy in the world.

Could life expectancy increase another 30 years in the 21st century, making centenarians the norm? Some researchers are skeptical. They point out that another 30-year gain in life expectancy would require a precipitous drop in death rates among older age groups. Accomplishing this would necessitate vanquishing cancer, heart disease, and a host of other complex illnesses, a feat many consider unlikely. But others are optimistic. As scientists unravel Methuselah's secrets, they believe we will identify targets for life-prolonging medicines. The optimists predict that these discoveries will eventually extend both the average and the maximum human life spans. The 20th century's oldest person lived just over 122 years. Some believe that this century's record holder will reach 130—or even 150.

THE ANTI-AGING DIET

Scientists can't yet transform 70 into middle age for us, but they've done the equivalent for other organisms by putting them on a diet. Give a worm, a yeast, a mouse, or a fruit fly a nutritiously sound but calorie-restricted diet, and it will live up to 40 percent longer than amply fed members of its species. What's more, it will do so in good health—at least within the protective cocoon of the lab. No one has tested how these organisms might fare under more stressful conditions. Thus far, however, the only ill effect of the diets, which reduce normal calorie intake by 30 percent, is reduced fertility. Researchers speculate that, when food is scarce, the body redirects its energies from reproduction to cellular main-

tenance and repair so that it can wait out the tough times, postponing reproduction for a more propitious moment. This has the effect of postponing aging as well.

Researchers are now beginning to understand how the body keeps itself youthful when calories are few. They hypothesize that a gene called SIR2 sets the pace of cellular aging in response to the cell's metabolic rate. Equivalent versions of this gene appear in organisms ranging from bacteria to humans.

LEVELS OF NAD, *the chemical shown above that is produced by a gene, appear to be the signal that yeast and roundworms use to set their rate of cellular aging. Researchers believe that NAD may play such a role in humans as well.*

Leonard Guarente and colleagues at the Massachusetts Institute of Technology, Cambridge, the United States, are illuminating how SIR2 works. They've discovered that the protein it produces can't function alone. It needs a helper chemical called NAD. But NAD has

another role in the cell as well: It is essential for converting food into energy. When food is plentiful, the cell preferentially uses NAD for metabolism, leaving little over for SIR2. When food is scarce, SIR2 acquires the NAD it needs to become active, and its actions slow aging.

Guarente has demonstrated the gene's anti-aging action in yeast and also in a worm called *C. elegans,* leading him to deduce that SIR2 is a universal regulator of aging. Intriguingly, SIR2 works its life-extending magic differently in the two species. In yeast, SIR2 slows the hands of time by silencing any unneeded genes, thereby preventing them from doing damage. In *C. elegans,* it mutes the effects of insulin, thereby slowing growth and, with it, aging. How—and if—SIR2 can modulate aging in humans is yet to be discovered.

RADICALS ON A RAMPAGE

The slippery little *C. elegans* has confirmed the role of another suspected agent of aging, too—oxygen. As cells convert oxygen and food into energy, they release highly reactive molecules called free radicals. Like a hyperactive toddler let loose in the living room, free radicals bound about the cell, slamming into proteins, fats, and DNA. The body protects itself against this bombardment by producing enzymes that convert the free radicals into harmless chemicals and by repairing any damage done. Over time, though, the body's defenses gradually weaken and damage accumulates.

FRUIT FLIES *with a mutated gene dubbed INDY, for "I'm not dead yet," live twice as long as normal. The mutation seems to impede the flow of nutrients into cells and may explain why low-cal diets extend life for yeast, mice, and other organisms.*

Scientists have suspected that free radical damage contributes to the changes we recognize as aging. Studies in *C. elegans* suggest they are right. Worms given drugs called antioxidants, which protect against free radicals, live 50 percent longer. Genetic studies provide further evidence. Several mutations correlated with longevity in worms, fruit flies, and mice also confer resistance to oxidative damage.

A towering caution sign flashes above these findings on aging, however. None of the age-busters—not calorie restriction, not SIR2, not subdued free radicals—has been shown to operate in humans. Because they influence aging in so many other species, however, many researchers suspect that they do in humans, too. But they still won't explain all age-related changes. Although some of the single-gene mutations that bestow long life in laboratory animals increase resistance to oxidative stress, not all do. As yet unidentified mechanisms must be involved. What's more, scientists have identified additional, specific influences on aging. They know, for example, that sugars can grab onto proteins in our tissues, causing them to harden. The resulting compounds are appropriately called AGEs, for advanced glycation end products. AGEs yellow teeth, discolor skin, and, worse, stiffen arteries and cloud the eye. Odd goings on at the tail-ends of chromosomes (see box), the flux of hormones, and cellular damage unrelated to free radicals may also turn our minds and bodies frail.

BOTTLED LIFE

The wealth of findings on aging casts doubt on the long-held notion that nature places an absolute limit on human longevity. Rather, a handful of malleable, if not reversible, processes most likely contribute to our decline. As researchers identify these processes and learn to manipulate them, they may be able to forestall the weary hearts, forgetful brains, and other changes of old age.

If calorie restriction turns out to guarantee a longer, healthier life, for instance, tomorrow's scientists may be able to develop a drug that stimulates the cell's response to semi-starvation,

enabling us to have our cake and live longer, too. A potion that mops up free radicals or pulverizes the sugar-protein bonds that form AGEs might similarly extend vigor, health, and years. Indeed, a potent and promising antioxidant is in the early stages of testing, while an "AGE breaker" that restores elasticity to stiffened tissues in animals is already in clinic trials. The AGE breaker has rejuvenated hearts and arteries in old dogs and monkeys. Should it prove safe and effective for people, it may become available just as the baby boomers swell the ranks of the retired.

No future treatment is likely to provide a perpetual fountain of youth, but that may be just as well. Even a 10-year increase in the average life span would effectively force society to reconsider the roles and resources it extends to its seniors. Besides, if life lasted forever, would we still love it so?

TELOMERE TUNE-UPS

The tail ends of chromosomes, called telomeres, act as a biological stopwatch, counting down the lives of a cell. They grow shorter with each cell division because the mechanism that copies our DNA cannot copy the last bits of the telomeres. In the lab, having too many critically short telomeres pushes cells into senescence. The cells can no longer divide and begin to churn out a different set of proteins than in their youth, thus altering their actions. Whether cells within the body reach senescence too is not yet known. But if they do, their altered functioning could contribute to aging in tissues that renew themselves, such as the skin and blood vessel linings.

Restoring chromosomes to their youthful lengths might thus reverse some age-related changes. By infusing human cells with an enzyme that stimulates telomere growth, Jerry Shay and colleagues at the University of Texas Southwestern Medical Center, Dallas, in the United States, have shown that rejuvenating chromosomes is indeed possible. They are now investigating treatments based on this principle. If a patient's own cells can be harvested—"telomerized"—and then returned to the body without causing cancer, this approach might one day counter age-related diseases such as macular degeneration or arteriosclerosis. More people might enjoy a healthier—and perhaps even longer—old age.

AMERICAN LOBSTERS (above) and other long-lived species have very lengthy telomeres (red, below). Telomeres are the ends of chromosomes and they grow shorter each time a cell divides. Researchers believe they may play a role in aging.

The Persistence of Plague

In all of human history, we have eradicated only one disease: smallpox. As new scourges arise and old ones rebound, infectious diseases threaten the promise of ever-improving health.

In a remote mountain village in the Dominican Republic, in the year 2000, something that wasn't supposed to happen did: Polio made a comeback. Supposedly eliminated from the Western Hemisphere almost a decade before, the disease crippled six children in the Dominican Republic and one in Haiti. This small but sobering outbreak was one more in a mounting list of reminders that microbes are not done with us yet.

According to a United States government report, 20 old foes, including tuberculosis (TB), malaria, and cholera, have reemerged or spread geographically in the last quarter of a century. TB, for one, roared back in the 1980s and 1990s, killing over 1.5 million people in 1998 alone. It's taken hold in the overcrowded slums of poor cities and in the homeless shelters and prisons of wealthier ones. It is epidemic among AIDS patients worldwide. Many TB victims have new, multidrug-resistant strains of the disease and will die despite treatment.

TB is far from alone in defeating the very drugs that once conquered it. Malaria, pneumonia, meningitis, gonorrhea, salmonella, strep, and even bubonic plague have appeared in drug-resistant forms. Since the mid-1970s, another 30 newly identified and as yet incurable killers—among them AIDS, Ebola, and West Nile virus—have added to the rising tide of infectious disease, a tide that threatens us all.

MOBILE MICROBES

Today, microbes travel at the speed of a jumbo jet, easily crossing oceans before revealing themselves in coughs or fevers or chills. The strawberries that brighten Northern Hemisphere supermarket aisles in February, the traveler who paddles the remote stretches of the Amazon, the refugee seeking safer shores—all can unwittingly carry pestilence from port to port. In so mobile a world, a pox upon any nation is potentially a pox upon one's own.

Unfortunately, health officials who sound this warning may be this

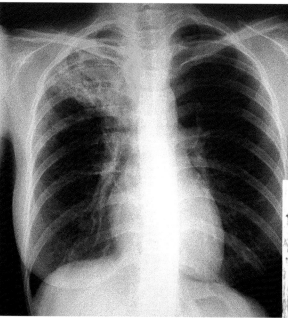

THE BATTLE *between science and infectious microbes is far from over. TB, a disease caused by the bacteria* Mycobacterium tuberculosis, *has become resistant to many of the drugs that stopped its rampage last century. The chest X-ray of a 31-year-old female patient (above) shows the upper left lobe of the lung colored red, with affected areas in green. Pneumonic plague, a deadly epidemic disease, resurfaced among the people of a Madagascan village (right) in 1983.*

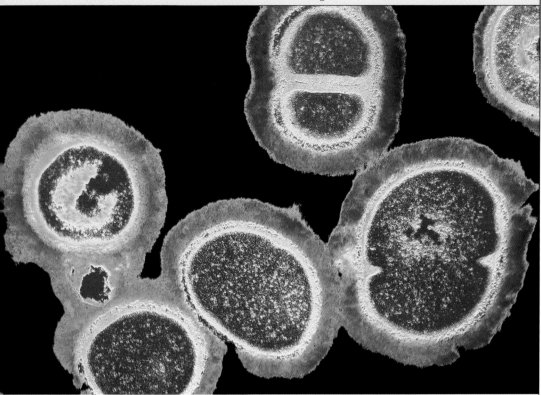

century's Cassandras, appealing to governments that pay little heed. Infectious diseases take their greatest toll in the poorest nations because of persistent poverty, poor sanitation, malnutrition, overcrowding, inadequate health-care systems, and war. Whether rich nations will devote enough resources to improve health in those countries is an open question. So, too, is whether the citizens of wealthier nations have the resolve to address threats to future health within their own borders, where profligate use of antibiotics threatens to catapult us into a future frighteningly like our disease-ridden past.

BREEDING SUPERBUGS

Only around half of the two million pounds (907,000 kg) of antibiotics produced in the United States each year is used to treat human disease—and half of that is used inappropriately. Patients don't finish their medications or they demand, and doctors write, prescriptions to treat viral infections impervious to the pills.

Veterinarians use some of the remaining antibiotics to cure animals, but most of it goes to doctor our food. We dine on chickens that have been fed antibiotics to promote growth and on apples sprayed with them to produce picture-perfect fruit. We add other germ killers to soaps, shower curtains, highchairs, and even toys, thinking we are creating a healthier environment for our children and for ourselves. But we are not. We are actually creating an army of superbugs. Our germ-repellent environment

THIS COLORED MICROGRAPH *shows a deadly cluster of* MRSA (Methicillin-resistant Staphylococcus aureus) *dividing. This bacteria is resistant to most antibiotics and is increasingly common in hospitals, infecting the wounds of patients.*

kills susceptible strains while leaving the more drug-resistant ones to flourish.

The problem of drug resistance is huge in hospitals, where resistant strains of three of the most common bacteria have emerged. These bugs are now all but invincible—only one known antibiotic still defeats them, often with some difficulty. Should bacteria develop complete resistance to this drug, the 100,000 to 150,000 people who die in the United States each year as the result of hospital-acquired infections may soon seem like a pittance.

Reversing the rise of infectious diseases is in our power, however. We can restrict our use of antibiotics and antibacterials. We can deliver essential drugs to developing nations. We can establish worldwide surveillance efforts to quickly identify and contain outbreaks of both known and emerging diseases. We can also redouble our efforts to develop new antibiotics.

But we cannot be complacent. Microbes will continue to evolve. They will swap genes, giving rise to newly resistant strains. They will mutate. They will jump to humans from other species. Albert Camus observed in *The Plague* that pestilence never dies. Once gone, it returns to "rouse up its rats again and send them forth to die in a happy city." He was right.

In the end, our society will be defined not only by what we create but by what we refuse to destroy.

Conservation by Design,
STEVE McCORMICK, The Nature Conservancy

CHAPTER TWO

SOCIETY

The Future of Work

At least half the jobs people do now did not exist 25 years ago

and at least one-third of the jobs that today's university

entrants will fill have not yet been invented.

For most of us, despite the many dazzling advances that technology will continue to deliver, paid work will remain one of the few constants of our adult existence.

But like so much else in the 21st century, the way we work —where, how, and for whom we do it, how long we spend doing it, even how many years we commit to working—is a concept that is changing. Indeed, some people now argue that change itself is becoming the business of our times.

The relentless momentum of technological intelligence, which sees computer power double every 18 months, has already erased whole categories of jobs that had been economic mainstays since the advent of agriculture 10,000 years ago and since the productivity explosion of the Industrial Revolution in the nineteenth century. In Western Europe at the start of the 21st century, only 5 percent of the workforce was engaged in agribusiness. Massive computer-controlled machines on consolidated farm lots now do the work; that is, the watering, fertilizing, harvesting, and packaging of primary produce.

The centralization of manufacturing fueled the sudden and massive growth of great industrial cities, providing employment for manual laborers and leading to the rise of the union movement. Currently, both are in decline, with "dirty jobs" disappearing because low-skilled workers are being replaced by ever-more-sophisticated machinery and robots that operate around the clock. The collective clout of labor is being progressively weakened to the extent that some Western economies only have around 10 percent union membership.

THE SELF-RELIANT WORKER

When it was first invented in 1975, the personal computer (PC) had about 2,300 transistors

BY 2020 *the office of the future may contain only one or two core workers (left). For most workers, the attainment of their degree (right) will be the beginning of a lifelong commitment to new learning.*

on a chip. The PCs of today have 7.5 million transistors on a chip and in another decade, that computing power will have been magnified exponentially. In the race to keep up with the evolution of what has become the primary tool in so many enterprises, future workers will be forced to be ever more mobile, adaptive, and conscious of the need to continually upgrade their competencies.

As employees become more self-reliant, so they will become increasingly aware of their value as individuals. The one-company man or woman, who in the past expected to occupy a position doing almost the same thing for three or four decades, is evolving into a multiskilled, highly responsive professional who can wear many hats, depending on the project at hand.

This person might be a "core worker" retained at company headquarters in a vital role in return for a personally tailored package of pay, car, salary, and share and pension incentives. But the better the remuneration, the more "languages" that employee will have mastered in terms of the company business.

That worker may be an accountant with sound knowledge of human resource management; or an information technology (IT) expert who can also manage the explosion of potential e-business opportunities. He or she may be an engineer who can double as an information analyst and sort through the relentless stream of incoming data and organize it as relevant or non-relevant. If that employee is highly skilled, he or she will be seen by the company as talent to be retained, but loyalty can no longer be presumed. In the age of

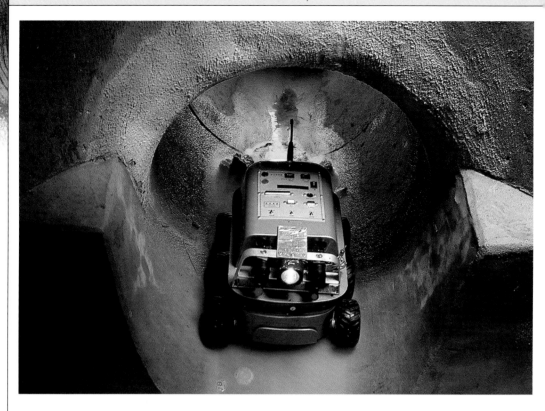

DIRTY JOBS *are disappearing as ever-more sophisticated robots like Kurt (above) take over. It is thought that eventually similar robots will inhabit, inspect, and maintain the sewer systems in cities such as Bonn, Germany.*

workplace individualism, already shown by the fact that about 15 percent of professionals voluntarily change jobs each year, good workers can career surf at a whim.

ONE WORKER—MANY CAREERS

Where once a work history denoting too many rapid changes caused uneasiness for prospective employers, it is now evidence of a worker who is keeping abreast of change.

In Silicon Valley, heartland of the IT industry that is driving globalization, spending more than two years in a job is considered career suicide. And in that Californian hothouse, some 30 percent of employees leave each year to get a better deal, or to work for themselves as consultants who may boast no specialized job description on their business cards.

Such highly mobile workers have, however, begun to attract interesting new classifications. They are labeled "hunter-gatherers." In the IT industry, they can be confident of getting work anywhere in the world because until at least 2010, the need for computer professionals will far exceed the numbers that can be trained. In the future workplace, in which e-commerce is expected to become dominant, IT literacy will be the absolute prerequisite for most positions.

In the broader workplace of the Western Hemisphere, too, low unemployment figures have combined with other factors to produce another new phenomenon of casual, part-time, or contract workers. In Australia, this trend towards "casualization" already accounts for almost half of the paid workforce.

These workers take up jobs that have been outsourced from corporations and they, too, are expected to be multi-faceted in terms of the services they provide. Defined as "portfolio workers," they may appear occasionally in head office and be assigned to a temporary desk by an "office concierge." More likely, they will operate remotely from an external or home office into a corporation that might be located anywhere in the world.

In parts of Europe today, almost 25 percent of workers telecommute; it is predicted that by 2020, with ever-improving virtual facilities, connections, and cost efficiencies, just under 50 percent of workers will not leave home to check in at their workstation. Virtual offices can be set up wherever the employee happens to be. However, even with a fantastic vision of virtual workstations that can be worn on the body like a piece of clothing, futurists predict that centralized bricks and mortar corporate offices will still exist. What will change will be their scale and function.

Offices will no longer be primarily dedicated to the processing of tasks but rather they will be places of intense real-time interaction with other human workers. The faceless virtual world is recognized as highly efficient but potentially lonely, so increasingly in the future offices may serve more as venues for socializing between individuals engaged in similar business endeavors.

THE END OF NINE-TO-FIVE

Outsourcing, economic globalization, and consumer demand for 24-hour convenience (already evident in more flexible shopping hours) are ringing the death knell of the traditional nine-to-five work pattern. Work now takes place 24 hours a day, seven days a week. One of the fastest growing industry sectors, the $1-trillion-a-year worldwide operation of call centers that centralize the customer service needs of countless firms, has workers clocking in at all hours.

Flexibility is the expectation of these new-age employees and, if the pundits are correct, quite soon their services will be streamlined enough to spearhead a move toward a standardized six-hour working day. Until then, however, those who work long and hard in the city and who have little time available for leisure will be calling on another outsourced service sector that is already demonstrating a massive growth curve.

THE FEMINIZATION OF SERVICE

Household service—comprising workers such as house cleaners, shoppers, and those who mow the lawn, clean the pool, walk and wash the dog, or provide hands-on labor for tasks that can't be dealt with in a virtual reality setting—will continue to be a booming sector. As with the customer service and retail industries, it is being largely fueled by the number of women in the workforce.

In some Western countries women now outnumber men in university courses, and at least half of all new small businesses are being started by women. This trend is predicted to continue as birthrates continue to fall. But even with the feminization of many industry and service sectors, the plummeting birth rate and a concurrently aging demographic (in Europe by 2020, about half the workforce will be over the age of 50) will result in a need to retain older workers. The age of retirement may also have

A BOOM IN PRACTICAL, *hands-on labor such as gardening, shopping, cleaning, and dog walking is expected to continue as more and more women enter the workforce.*

to be waived as older workers retrain for new jobs in the economies of the future.

Undoubtedly the nature of our work will change, as will the age and mobility of the workforce, but like the motive of being paid, there is one other constant that will characterize the workplace of tomorrow. To stay relevant, workers will need to accept that formal education no longer ceases with the attainment of a single certificate or degree. Ongoing training and a lifelong commitment to new learning are part of the package in the dawning world of work. The key to future employment will be the ability to meet the changes with a constantly evolving set of skills.

JOBS' WORTH

In just one decade, hundreds of positions have gone the way of the village blacksmith. They include typists, subway drivers, mail carriers, lift drivers, and many types of manual and process workers. Also disappearing fast are bank tellers, traveling salespeople, and middle managers with a singular specialization.

The Home of the Future

The home of the future will be technologically advanced,

energy efficient, and kinder to the environment,

while housework may become a thing of the past.

With land for urban development becoming increasingly scarce, the trend is expected to move away from houses toward apartments, with the home of the future being designed as a smaller dwelling accommodating more people. Multi-use areas will replace separate rooms with self-contained functions, so that a large, open area could be transformed from a living area during the day, to a dining area in the early evening, to a bedroom at night, by using various fold-away components mounted in walls, floors, and ceilings. At the same time, self-contained units, with bathroom and kitchen facilities, will become more common in the family home—the perfect and affordable housing addition for aging parents or older children.

DESIGN AND CONSTRUCTION

Simple design ideas will harness the sun, local winds, and land and house aspect to passively heat and cool a house. Clever designs will provide movable walls so that the house's shape can be changed according to climatic conditions. A living area could be opened up during the summer, then reduced in size to help with heating during the winter.

Alternative building materials, such as fiberboards (made from recycled materials), concrete, and plantation-grown softwoods, will replace the slow-growing hardwood timbers used in many contemporary houses. Straw bales, molded plastics, and mud bricks may also be used more frequently; these unusual materials place fewer demands on natural resources and reduce energy use because of good thermal insulation properties.

THE POWERED HOME

Future homes could actually become energy generators. In several current trials, houses with solar panels or other small household generators have been connected to the electricity grid. During times when the household's energy demands are heavy, the house draws the extra energy it needs from the grid, but when demand is low, the house creates an excess of energy which is fed back into the grid and provides the householder with an energy bonus.

Small fuel cells no bigger than an office photocopier could provide the energy needs for an entire household. These fuel cells, powered by burning methanol, break water down into hydrogen and oxygen; when this is recombined to form water, it generates electricity—just like a battery that never runs flat. While this technology is still expensive, work on cheaper versions could make it widely available.

THE SMART HOME

Digital technology will be integrated into future homes to help manage the comfort of the occupants and connect the home to the neighborhood and the world. Conduits built into the walls will carry information from sensors, which monitor temperature, light levels, and other conditions in each room, to a central control unit. This unit will then automatically modulate conditions throughout the house to match those set by the occupants—turning lights on and off, adjusting air conditioners and heaters. It could also keep track of who is in the house and where they are, making adjustments for each person. With such tracking capabilities, it could double as a security guard.

Once it is possible to connect the central control units to the Internet, a host of automated tasks will become possible. A smart refrigerator will not only record what goes into and comes out of it and generate a shopping list when drink and food items run low, it will then order the groceries needed via the Internet for home delivery.

DOMESTIC ROBOTS *could eventually act as security guards, patroling the interior of a house (left). Already on the market are "intelligent" refrigerators. The display (below) provides reminders of food items that may have been forgotten about and alerts users when food is about to reach its "best before" date.*

HOME HELP

Various home helpers may soon take the hard work out of housework. Existing smart appliances range from digital clocks that reset themselves after a power shortage to microwave ovens that download recipes from the Internet and self-diagnose mechanical problems. Many common appliances are being adapted for remote control use via telephones or the Internet. New smart appliances are also being developed, such as the smart vacuum cleaner, which will power itself along the floor and clean the whole house without running into furniture or falling down the stairs.

Simple domestic robots are already available, and those that can do all household chores are not very far away. For the time being, their owners will be in control of them; the fully autonomous house robot beloved of movies and fiction is still a dream of the future. But eventually, our homes may well become places where all we will need to do is program the central control unit, the robot, and the appliances, and then sit back and relax.

THE HOME OF THE FUTURE *could fast become many people's office of the future, with the space and comfort, adaptable rooms, wireless technology, and intelligent appliances it promises.*

Genetically Modified Foods

Genetic engineering offers the promise of food products
we could previously only dream about. But will the public
eat food that has been engineered in a laboratory?

Designing food to meet our desires may sound like science fantasy, but the final decade of the 20th century saw the rise of gene technology—the ability to manipulate the make-up of an organism to suit our needs. For some, this technology will be our salvation, allowing us to create the food we need to feed an ever-growing population. For others, it is toying with the fiber of life itself, with many unforeseeable consequences.

The blueprint of life is deoxyribonucleic acid (DNA), which is organized into strands. Sections of strands are called genes and these genes provide a code for the specific features of an animal or plant: its color, its shape, its ability to make a protein, and so on. An organism's genome is the total of this code, the sum of all its genes.

AT A RESEARCH CENTER *in Israel, a syringe injects DNA into fish eggs (below). The ever-increasing consumption of fish worldwide has combined with the very serious issue of dwindling fish stocks to prompt a great deal of research into genetically modified fish.*

Genetic engineering offers the ability to change and restructure the genes that build animals and plants. With it, we can move genes around within an organism or take a gene from one organism and put it into another. We can also turn genes on or off, to start or stop them from doing their job and producing a particular feature within an organism.

It has been argued that we have been genetically modifying organisms since we started selectively breeding livestock and crops. While this is in some ways true, it oversimplifies the power of genetic engineering. Selective breeding allows us to improve or manipulate the qualities of an animal or plant by selecting desirable qualities over several generations. For example, bigger pigs were produced by allowing only the largest pigs of each generation to reproduce. The desired qualities must come from within the species, and the results take several generations to become apparent.

Genetic engineering, however, allows us to introduce traits from any living organism, not just from members of the same or closely

related species—thus genes from fish have been placed in vegetables and crops. And the desired results are produced within a single generation. In short, genetic engineering is far more precise than selective breeding; it can draw from an infinitely greater palette of desired qualities, and it is much faster.

GENETICALLY MODIFIED ORGANISMS

A range of genetically modified organisms (GMOs) is already available or in development. To date, the majority of these genetic manipulations have been made to benefit food production techniques. For example, crops have had genes inserted that produce insecticides, giving them automatic protection from insect attack. A tomato has been engineered to contain a gene from a fish to help it survive for longer in cool-storage. And salmon has been genetically modified to mature in less than half the natural time, dramatically cutting production costs.

However, there is a subtle difference in the GMOs currently being developed. The new GMOs are being enhanced for the benefit of the consumer rather than the producer—coffee that has no caffeine, vegetables and fruits with increased vitamin and nutrient content—and modifications are also more environmentally sensitive. This change in philosophy by the GMO manufacturers has been in response to a consumer backlash against the perceived corporatization of agriculture.

GMOs ENHANCED *for the benefit of the consumer, like coffee beans that contain no caffeine (above) show a change in philosophy of GMO manufacturers. Opponents argue that, with a direct link between GM crops and the death of monarch butterflies (top), environmental improvements don't go far enough.*

THE DEBATE

Proponents of GMOs argue that such crops will increase the productivity of our farmlands and the quality of our foods. We cannot at present produce enough food to feed everyone on Earth, and the availability of arable land continues to decrease. High-yielding GMOs could be engineered to grow in poor soils and marginal climates to help solve these problems. Proponents also argue that GMOs are safe to consume: There is already a small quantity of DNA in most foods that we eat and drink,

69

and changing the form of that DNA will not make it, or the foods, dangerous.

Opponents argue that we could be engineering super-weeds or critically rearranging the ecology of our farmlands and destabilizing the balance of nature. Some studies revealed that birds, insects, and other organisms are disappearing from areas where genetically modified crops are grown. Another study showed that monarch butterflies died when exposed to pollen from plants that had been genetically modified to produce insecticide. And while there is no danger from eating DNA that has been added to an organism, it is theoretically possible that such additions could lead to the production of toxins in that organism. However, there is no evidence of this to date.

There are a great many ethical and political concerns surrounding GMOs, too. Some communities view the direct manipulation of DNA as an affront to their way of life, while others consider it a gift that should be utilized. Complex questions surround dietary taboos. For example, would it be wrong for a Muslim or a Jew to eat food that contains genes from a pig?

With regard to the economics of GMOs, gene technology is extremely expensive and only a handful of multinational companies are capable of funding it. Growing a GMO crop usually locks the farmer into purchasing additional products from the same company that developed and sold the GMO seed. For example, a particular type of cotton is resistant only to a pesticide produced by the same company that manufactures the pesticide-resistant cotton seed, so the farmer who grows that cotton is obliged to buy the company's pesticide as well. This corporate approach to agriculture creates a perception that genetic engineering is not a particularly benevolent tech-

OVER HALF the soya beans grown in the United States are genetically modified. The arrival of non-GM soya from Brazil in Montoir Port in France (above) is in response to a French supermarket chain's settlement of a soya network with no GMOs.

nology. Complex questions regarding the ownership of a GMO crop also arise: Is it the property of the farmer or the seed manufacturer? And some opponents believe that an increased reliance on expensive GMOs will only increase debt in the developing world. In an effort to combat this, there is a global push for GMO manufacturers to give their products to such countries for free, or for a nominal fee.

PUBLIC REACTION

The reaction of the general public to GMOs has varied around the world. Many European countries are adopting a cautious approach to the introduction of GMOs into the environment and the food chain, although in several places, negative feelings towards GMOs are so strong that protesters have ripped up fields of genetically modified crops. The United States was initially more accepting of GMOs, but the consumer backlash there is now similar in scope and energy to that found in Europe.

There is increasing pressure from all corners to identify foods that contain GMOs through the detailed labeling of food products.

ACTIVISTS HAVE WAGED a war against genetically modified crops and products that contain them. This Canadian protestor, dressed as a mutant tomato on Parliament Hill in Ottawa, is demonstrating for the mandatory labeling of genetically modified foods.

A number of companies have promoted their food products as being free from GMOs and this has become an effective marketing tool.

Countries of the developing world, on the other hand, have generally been enthusiastic about the prospect of growing GMOs. Often with large populations but increasingly limited fertile arable land, they see GMOs as a possible solution to starvation and famine, as well as an economic boon—a way of increasing production to a point where they can export food to receive income. However, the reluctance of countries in the developed world to deal with crops contaminated with GMOs is severely restricting the introduction of GMOs into the developing world.

THE FUTURE

The future of GMOs as food depends on the public and consumer reactions to the debate currently under way around the world. If sufficient numbers of consumers deliberately avoid genetically modified food for whatever reason, the profit motive behind producing GMOs will be removed and GMO products will disappear from supermarket shelves. There is already some evidence that this is happening.

However, if the message that genetically modified food is safe to eat is accepted, then this technology promises an ever-expanding list of new products, and it will be embraced because of the benefits it offers in food quality and quantity—and the potential gains for human health worldwide. The genetically modified food of the future would be vitamin- and nutrient-enriched, with the possibility of vaccines being introduced to the food to provide protection against various diseases.

The future environmental impacts of GMOs are similarly uncertain—as are the political and economic impacts—and how they will be dealt with by farmers, land managers, GMO manufacturers, and governments is unknown. It may be found that, while growing GMO crops is profitable, the longer-term effects on the environment could be devastating and cost more than the short-term profits could justify. Any impacts upon biodiversity and land quality are currently unknown and will therefore have to be addressed as they occur; which may severely limit the number and type of GMOs produced.

As with any new technology, there will be associated costs and benefits. But while we may now know what benefits genetic technology promises, we have only a vague inkling of what the potential costs may be.

GENETIC MODIFICATIONS *can create sturdier plants—by making crops more tolerant to adverse conditions, by making crops that are resistant to herbicides and pesticides, or by making crops poisonous to common pests. The end result is a crop that's beneficial to producers. But is it beneficial to consumers?*

Sustainable Agriculture

The 20th century saw a huge boom in food production, but at the cost of severe land degradation and environmental destruction. How will we feed ourselves in the future?

The Natufian people of Palestine, when they made the world's first crude sickles to harvest wild wheat nearly 10,000 years ago, could not have known they were starting a revolution that has come full circle.

Agriculture as a formal science was born just 200 years ago. The Industrial Revolution provided the mechanized means to sow and harvest crops, and to fertilize and protect them with pesticides. After World War II and some extensive tree-clearing, food crops were grown in vast monocultures (single species). Animals were farmed intensively in the same way. This saw food production triple between 1950 and 1980, while food prices dropped by 25 percent.

We are now discovering that this intensive farming severely depletes land fertility, damages soil structure, and causes erosion and salinity. Vast areas of agricultural land are being exploited to their limits; in developing countries, the pressure to grow food for expanding populations has left many areas devastated, and new farming land is becoming scarce. Meanwhile,

world population is expected to increase by another 1.5 billion people by 2020.

The challenge for agricultural science is to find ways to produce enough food—and to do it sustainably. This can be achieved by limiting the area used for growing food, and even producing food without using land. Farming methods need to protect soil and ecosystems and reduce water usage, while retaining more tree cover in food production areas.

MORE FOOD, LESS LAND

One way to limit land use is to cultivate high-yield strains so that more food can be grown in a smaller area. A strain of corn developed in Mexico that uses traditional breeding methods contains twice as much protein as existing varieties, and there are plans to cultivate this Quality Protein Maize in Africa, Asia, and Latin America.

Genetically modified organisms (GMOs) are not widely regarded as a tool for sustainable agriculture, but some GMOs might help to

INTENSIVE FARMING (left) involves clearing vast tracts of land to prepare it for one type of crop. This method only works in the short term as the soil diminishes in fertility and suffers from increased erosion. These carrots (right) were grown organically, a method where a number of different types of produce are grown on a plot without the use of pesticides.

reduce the need for damaging agricultural practices. For example, plants engineered to be resistant to root damage from insects would remove the need for intensive plowing, which can cause soil erosion and water loss. Unfortunately, the agricultural biotechnology companies that develop GMOs hold strict patents, often making such technology unaffordable to poorer countries. However, these companies have been called upon to provide appropriate GMOs at a lower cost, or without charge, to such countries; one United States company has already done this with its patented rice strain.

Land use could also be limited if we moved towards a predominantly vegetarian diet, for the simple reason that it takes four times the land area to produce the same amount of beef as grain. However, given that limiting their meat intake is not something that many people consider, meat would need to become much more expensive before consumers would think of changing to a more vegetarian diet.

Some day, pressure upon agricultural land may be reduced by growing food in the laboratory. In Australia, a type of algae containing omega fatty acids has been developed and is included in some commercially available breads

and nutritional supplements. Laboratory-grown food is likely to become a cheap way to mass-produce certain dietary essentials.

FARMING METHODS

Many Third World agricultural experts argue that the only way to ensure sustainable agricultural production is to use traditional organic farming methods. Such methods favor multiple crops rather than a monoculture, use no chemical pesticides or fertilizers, and do not disrupt the soil structure. Organic produce is usually more expensive, and its often less-than-perfect appearance is still unappealing to many. If the small but growing demand continues, however, it will motivate the mainstream agricultural companies to grow more food organically.

Currently, such companies view traditional methods of farming as too labor intensive, expensive, inefficient, and unproductive when compared with modern, high-intensity mono-culture methods. Yet studies have shown that chemical-industrial farming requires up to 60 times the energy input (from fertilizers and fuels) of an organic farm growing mixed crops to produce the same amount of food. Farmers in Yunnan Province, China, are significantly increasing their yield of rice by growing different strains together. The most desirable—sticky rice—is less prone to fungal disease when grown next to more resistant strains.

Embracing a sustainable system of agriculture will require a radical shift in thinking around the world. But if costs can be kept down, and as benefits become more apparent, sustainable agricultural methods will attract greater interest from both farmers and agribusiness.

SOIL DEGRADATION affects many plants not on designated farmland. Forest clearance for farming elsewhere creates erosion and salinity problems for trees in the wilderness (left).

Finding New Resources

The steady consumption of known ore resources and exploding demand for minerals and earth materials that can change the entire direction of technology are sending exploration geologists a clear message ... Find more!

New resources are continually needed to replace the mineral resources used by our spiraling industrial expansion. But exploration geologists and raw material manufacturers, working with an unprecedented wealth of know-how and technology, are also meeting demands for new "super metals" such as magnesium, silicon, and titanium, the backbone of the technological frontier.

From the creation of cyberspace to the mass production of the humble hypodermic needle, the 20th century was not just a triumph of technology, but also of materials. New metals, new alloys, and new manufacturing methods revolutionized the way in which we use our mineral resources. Despite such exotic developments, the global resources sector earned an "old economy" tag for its lack of change in the basic way it did things.

But it had succeeded in a gargantuan scale-up of its operations. In bulk mining, trucks grew to carry 400 tons (406 tonnes) and crane-like draglines weighing 4,000 tons (4,065 tonnes) could swing truckload-sized scoops of rock over 300 yards (275 m)—machines that could dig the Panama Canal in a month. Humans learned to literally shift mountains. A single mine could drink five million US gallons (19.3 million liters) of diesel in a year. The 20th century was one in which global consumption of coal and metal rocketed and the price manufacturers paid for metals steadily fell.

RESOURCE FRONTIERS

But can we mine without moving the mountain? In North Carolina, United States metallurgist Bill Drinkard has found ways to mine a deep vein of copper through a drill hole using a method he calls electrical mining. Drinkard is one of a new wave of "hydrometallurgists" whose work could finally make smelters—the energy-intensive furnaces that can be traced back to 5,000 BC—into a thing of the past.

In hydrometallurgy, leaching ores directly into the solution where they can be electroplated prevents harmful atmospheric emissions. Another bonus is that the ores need not be ground to a fine sand to release their metals, a highly energy-intensive practice.

We can also expect that low-cost underground excavation will make subterranean space an important new resource for storage tanks, parking, and transport. Porous geological formations vacated by oil and gas may become the reservoirs for carbon-rich fluids reinjected into the earth out of harm's way.

Mining will increasingly be done in total automation, leaving only machines to endure

20TH-CENTURY MINING *saw a huge scale-up of operations which included the development of gigantic machinery, such as this dragline bucket stacking oil shale at a Canadian mine.*

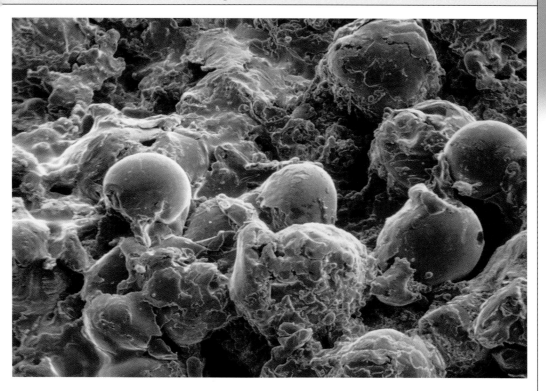

IN THE RECENT PAST, *the high-demand metals were zinc, lead, and copper, but now titanium (above), magnesium, and aluminum are increasingly in demand for the production of light, energy-efficient cars and other consumer durables.*

the rigors and dangers of underground life. Control room operators will be the only people to see 600-foot (180-m) long wall coal miners cutting out coal seams deep in the earth and progressively dropping the roof behind.

Undoubtedly, the shifts in our metal usage will accelerate, with magnesium, titanium, and aluminum becoming the preferred metals for building light, energy-efficient cars, and many other consumer items. Fuel cell builders are set to explode the demand for platinum and are expected to consume over half a million ounces (142,000 kg) by 2010.

Also this century, geophysicists are likely to reach their holy grail of a "glass Earth," using sophisticated electrical, magnetic, and gravity-sensing instruments to "see" through ¾ mile (1.2 km) of Earth's crust in their quest to find new resources.

Above all, improved technological know-how in processing new parts of Earth's crust in different ways, and in making new and old materials perform new

wonders, is likely to stave off the risk of any apocalyptic resource drought.

BACTERIAL MINERS
The midday heat of Western Australia's vast deserts is enough to make any miner stop to draw breath. But in a loaf-shaped heap of ore at the Mount Sholl mine, billions of bacteria are busy leaching nickel out of sulfide rocks. Green fluids are not the only things flowing

THE SILICON STORY

There can be much more to finding new resources than stubbing one's toe on a rock. Take, for example, quartz silica, or silicon dioxide—by far the most common compound in Earth's crust. The humble piece of quartz is usually the first mineral to find its way into the junior rockhound's box.

Unlocking the silicon metal from quartz and harnessing its extraordinary properties was one of the defining scientific acts of the last century. The silicon chip was born and its benefits will reverberate through the 21st century. Today, this microscopic roadmap of printed silicon wires simulates the human brain in ever more powerful ways and surrounds us in computers, wristwatches, and dozens of other appliances. The resource had always been there; it was technology that unleashed it.

The modern mobile telephone has 42 metals in it, including silicon. In the fast lane of mobile phone users, the contents of this sleek consumer item may seem nearly irrelevant. But the extraordinary eloquence that inventors and designers used to build such devices from a broad palette of materials refined from mineral resources is an indication of the complexity we can expect in 21st-century resource consumption.

at the plant; this pioneering operation has presented its first profits, along with some important pointers for metal refiners.

Of interest to the industry's analysts is the fact that all this is being done without the costly heat-pressure vessels and enormous running costs normally required. The busy bacteria penetrate the crushed ore chunks and oxidize the sulfides. This dissociates the nickel and sulfur, with energy resulting, which the bacteria use. The ions produced can then flow from the heap to the recovery pond.

The invisible hero in the Mt Sholl heap leach is called *Sulphobacillus thermosulphidoxidans*, a naturally occurring bacteria from quite ancient genetic stock. Its new career as a mining bug is through the operators making careful adjustment to the environment within the sprinkled heap of ore, such as access to air bringing in carbon dioxide and oxygen and in the trace levels of nitrogen, potassium, and phosphorus. Biological leaching of heaps is one of the world's most promising low-energy metallurgical processes.

THE OUTER FRONTIERS

In a bid to find living examples of the geological conditions that formed the world's major zinc–lead–copper ore bodies, researchers went to the volcanically active plate margins of the deep sea. There, they found hydrothermal mounds the size of office blocks. Their upper surfaces were forested with vents known as "black smokers," jetting their hot, cloudy, mineral-laden waters into the sea at a depth of 1-mile (1.6 km). Indeed, these minerals of exceptional purity were still forming. Tracking

them down had been made considerably easier by the plumes of cloudy water hanging above them, spanning many square miles (kms).

What researchers weren't expecting was that the grade of zinc, copper, and gold found in these "living laboratories" was commonly ten times that of the same minerals found in mines on land. Papua New Guinea granted the world's first exploration licenses over such deposits to an Australian company, Nautilus Minerals Corporation. This event and ongoing discoveries around the globe suggest that "sea-floor massive sulfide deposits" (as they are known) may be quite common and that this century could see mining move offshore, just as petroleum extraction did in the last.

How will they be mined? Nautilus envisages a mining tool comprising existing technologies, using jaws or rotating picks to chew the ore off. It would then crush and pump the ore vertically through a suspended pipe to the mine ship overhead. The ore would then be processed on the ship or on land. Significant benefits are expected to apply in marine mining that do not apply in existing mines on land. Nautilus says that the absence of overlying rock to excavate or penetrate, the high grades, soft ore, and

reduced waste would make the mining of its marine sulfides more profitable than alternative metal sources on land, and also less environmentally damaging.

MINES WHISTLING THROUGH SPACE

The International Space Station poses some interesting resource questions as to the materials that will be used for further space construction. The high cost of Earth rockets and fuel, basic materials such as paneling, and even life-giving water, places a heavy drag on space construction. Recognizing this, some entrepreneurial scientists are researching whether the materials needed can actually be mined in space, and whether abundant supplies of a stainless steel-like nickel–iron alloy obtained from asteroids might be the key.

Moons and asteroids are small space bodies that have much lower levels of gravity at their surface than Earth. Even small electric "slingshots" could fire their materials into neutral gravity space or low-Earth orbits, with no massive space shuttle required.

By analyzing the light spectrum reflected by these heavenly bodies and comparing it with rock samples from the Moon and meteorites, scientists have collated a long list of available materials that could be transformed by focusing the Sun's rays on them by means of lightweight solar ovens that use foil mirrors and

IN THE SEARCH *for new sources of minerals, scientists discovered "black smokers"—hydrothermal vents deep beneath the sea that jet out super-hot water laden with very high grades of gold, copper, and zinc. The mining of such minerals, now in the exploration stage, may be a profitable industry in the future.*

intense 24-hour solar energy. The material will then melt or bake into a new shape of suitable design.

The researchers are mulling over what they know of 1,100 known near-Earth asteroids and considering which would be ideal targets for capture. Steered over to a useful near-Earth orbit, an asteroid could become a vast stockpile of nickel–iron, clay materials, a high-hydrogen bitumen, water, and carbon.

The deep, dark craters of our Moon could give us water from ice and abundant ceramic and concrete materials for general construction. From both sources, life-sustaining water "ores" could be the first and the most economically attractive product. The speed with which we colonize space will govern the timing, but once we're out there, using such local materials will be a natural progressive step.

SPACE COULD BECOME *a new resource frontier if techniques can be developed to allow mining of asteroids such as Asteroid Ida (below). Asteroids could be useful sources of nickel-iron, cement materials, and water.*

Managing Waste

With mountains of urban and industrial rubbish piling up

around us and concerns escalating about toxic refuse, waste

management is a major challenge for the 21st century.

As any tabloid journalist who has rummaged through a celebrity's trash can knows, people's rubbish reveals a great deal about them. Archaeologists and anthropologists apply the same rationale when painstakingly sifting through the rubbish dumps of our forebears. So what do waste production and management practices reveal about the human species in the 21st century?

THE DISPOSABLE SOCIETY

The most conspicuous aspect of global waste production is the sheer amount of it. In the developed world, we now dispose of far more rubbish than ever before. The reason is not simply that the human population has grown to an unprecedented level. As individuals, we discard far more than our ancestors did. Per capita, the biggest offender is the United States, where each person produces an average of about 4.4 pounds (2 kg) of trash each day—up from 2.7 pounds (1.2 kg) in 1960. Australia comes a close, and equally inglorious, second.

Another key characteristic of our rubbish is that much of it outlasts the average human life-span. Aluminum cans take up to 400 years to break down. Glass bottles take a million years. And, as far as scientists can determine, plastic materials like polystyrene and polyvinyl chloride (also known as PVC or vinyl) may persist indefinitely in the environment.

Because of the massive quantities of urban and industrial waste now being produced globally and the limited biodegradability of much of it, landfill sites used by many of the world's towns and cities are fast running out.

THE THREE Rs

To a large extent the answer is embodied in the simple mantra "reduce, reuse, recycle."

Reduction will be the main priority of everyone concerned with the planet's waste problems—from engineers to environmental activists—well into the 21st century. We can expect to see, for example, many more initiatives like those taken by glass and bottle

manufacturers during the past 20 years to reduce packaging materials. Since the late 1970s, the weight of plastic and glass bottles has been cut by up to 25 percent and 30 percent respectively.

Widespread adoption of reuse strategies will require a major change in mindset for the highly disposable societies that became so entrenched in the developed world of the late 20th century. During the early part of the present century, we are likely to experience a significant shift back towards the manufacture

THE LIMITED BIODEGRADABILITY *of much urban and industrial waste is of great concern. Although these corn chips stacked in polystyrene bowls at a sporting stadium (above, left) will be consumed by spectators within minutes, their packaging will last indefinitely. So will the toxic waste of a nearby Mexican factory, which pours into the Rio Grande on the United States border with Mexico (above). Nearby communities have suffered from birth defects and deaths as a result of this pollution.*

of products that can be reused. The automotive industry has already taken major strides in this area, with up to 90 percent of all car parts now designed for reuse through a process known as "remanufacturing."

Recycling completes the holy trinity of optimal waste management and in many developed nations there has been a remarkably high uptake of this practice. In Japan, for example, recycling rates of aluminum cans increased from 41 percent in 1986 to 78 percent in 1999. In the same period, the rate for recycling of steel cans rose from 38 percent to almost 83 percent. During the coming decades, such figures will continue climbing and a wider variety of products will become suitable for recycling.

GAR-BARGES

During the 1980s, unsightly barges laden with rubbish became potent symbols of the planet's emerging waste crisis. The most renowned, the *Mobro*, spent 55 days in 1987 roaming the Atlantic looking to dump 3,100 tons (3,150 tonnes) of New York garbage. During a controversial 6,000-mile (9,650-km) journey watched by the world's media, it was spurned by five US states and three other countries. Eventually, the *Mobro*'s cargo was reduced to ash in a Brooklyn incinerator and dumped in a landfill back in New York.

In 1986 the *Khian Sea* embarked on an equally contentious, but less publicized, journey, carrying 15,000 tons (15,240 tonnes) of ash from incinerated waste from Philadelphia. The *Khian Sea* changed its name twice as ports in four continents rejected its load. In late 1987, it dumped almost one-third of its cargo on a Haitian beach before that country's government sent it, and its unwanted load, elsewhere. One year later, the ill-fated vessel arrived—*sans* ash—in Singapore. Years later, it was discovered that the Atlantic and Indian Oceans had become the dumping grounds for its noxious cargo.

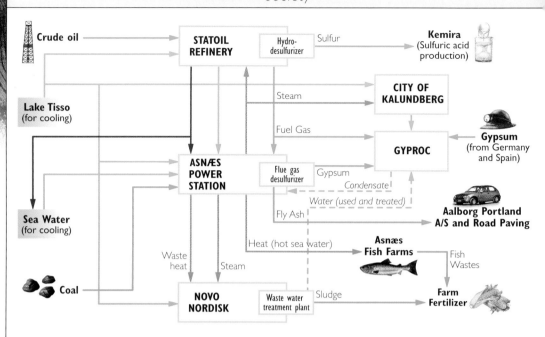

Crude oil — **STATOIL REFINERY** — Hydro-desulfurizer — Sulfur — **Kemira** (Sulfuric acid production)

Lake Tisso (for cooling)

Steam — **CITY OF KALUNDBERG**

Fuel Gas — **GYPROC** — Gypsum (from Germany and Spain)

ASNÆS POWER STATION — Flue gas desulfurizer — Gypsum

Condensate

Water (used and treated) — **Aalborg Portland A/S and Road Paving**

Fly Ash

Sea Water (for cooling)

Heat (hot sea water) — **Asnæs Fish Farms** — Fish Wastes

Waste heat — Steam

Coal — **NOVO NORDISK** — Waste water treatment plant — Sludge — **Farm Fertilizer**

Key to Chart

Water	————	Energy	————
Materials	————	Proposed	- - - -
Waste Water	————	Core Participant	▢

THE ULTIMATE GOAL

In nature, everything is ultimately reused or recycled so that there is zero waste. Many scientists who subscribe to the notion of industrial ecology believe that we should be setting a similar goal, but few urban or industrial settlements come close to this target. One that does is Kalundberg, a Danish town often cited as a paradigm for the future. Since the early 1970s, the philosophy in Kalundberg has been that one industry's waste is another's treasure. Today the town's industries and community operate much as a natural ecosystem does, according to a series of symbiotic interconnections. For example, the town's coal-fired power station channels excess steam through a district heating system supplying other local industries and residents. One of the beneficiaries, a biotechnology company, in turn provides a nearby farm with waste sludge as a fertilizer. The same company passes on excess yeast from a drug production process to local farmers for use as pig food. The power station also recovers sulfur dioxide from its chimney stacks and mixes it with calcium carbonate to produce gypsum, which is sent to a nearby plasterboard factory. These and many other such links keep waste and pollution in the town to the bare minimum.

TAMING TOXIC LEGACIES

There are many wastes, however, which cannot be reused or recycled. Either they are inherently hazardous to the health of living things, or they produce toxic by-products during treatment or disposal. The environmental group Greenpeace estimates that, each year, the industrialized world produces some 300 million tons

OVER THE PAST 20 YEARS, *the town of Kalundberg, in Denmark, has developed a web of materials and energy exchange among companies and the community (above). This "industrial symbiosis" has resulted in reduced costs, less waste, and environmental benefits. Meanwhile, in Sihanoukville, Cambodia, soldiers wearing protective clothing and masks head toward a 3,000-ton (3,048-tonne) pile of toxic waste (below).*

SPACE-RACE JUNK

Our waste production has not been confined to Earth. Billions of bits of old spacecraft and similar debris now orbit our planet, a legacy of a half-century of space exploration. Scientists track at least 10,000 of the largest, using Earth-bound radar and optical satellites. Up to 150,000 other objects, under four inches (10 cm) in diameter, can't be tracked. The rest are dust-particle size. Even small debris, however, is dangerous in space and a 1998 report by the National Research Council of the United States National Academy of Sciences warned of future risks to space craft and astronauts. Space agencies agree that the creation of space junk should be minimized, while acknowledging that a clean-up of junk already there would be too costly and technically problematic.

(305 million tonnes) of hazardous waste that comes under the first category.

For some forms of hazardous waste, bioremediation—the use of microorganisms and, to a lesser extent, plants to degrade or stabilize contaminants—is offering answers. Microbes have already been used successfully to clean up PCBs and petrochemical waste (including benzene) and to degrade other toxic challenges such as heavy metals and the pesticide DDT. Interest in and use of bioremediation has been growing steadily since the 1960s, and the tech-

nology now represents up to 10 percent of all pollution treatment. Many scientists and engineers believe, however, that what we have seen so far of microbes as toxic waste cleaners is just a glimpse of their true potential.

In some cases, there are simply no treatments available to neutralize or reduce the toxicity of hazardous waste. This is the case with the most insidious of all hazardous material discarded by the human species—high-level nuclear waste. Such waste comes mostly from nuclear power plants and decommissioned nuclear missiles. It includes highly radioactive forms of uranium, plutonium, and other elements. Some of these have a half-life (the time it takes for half of their atoms to undergo decay) of more than 100,000 years. Put simply, if this material had been around at the time our species first appeared on the planet, it would still pose a danger today.

Much of this waste is kept in short-term, on-site storage facilities at the numerous nuclear power stations operating around the planet. During the past decade, scientists and governments have been accelerating efforts to find safe, long-term storage solutions.

The most widely supported disposal option is burial in an underground repository designed to permanently isolate the waste from the surrounding environment. The waste would most likely be chemically stabilized and secured inside steel canisters before being entombed. Where to do this, however, is a major issue, because if ever there were a product that suffered from the NIMBY (Not In My Back Yard) phenomenon, it is nuclear waste.

Burning the waste in Accelerator-Driven Subcritical Nuclear Reactors, known as ADSNRs, is also a growing possibility.

Green Architecture

As we struggle to find ways to reduce our impact

on the environment, architecture has been identified

as one area where major gains can be made.

The homes, offices, and communities we live and work in are responsible for a great deal of environmental damage. For example, 50 percent of ozone-depleting chlorofluorocarbons (CFCs) produced in the world are used in buildings for refrigeration, air-conditioning, fire extinguishers, and insulation. About the same percentage of the world's fossil fuel is used up servicing buildings, and about one-third of the carbon dioxide emissions in the United States are a direct result of the building industry. In addition, a great deal of water is wasted during construction, and most urban developments are now designed to divert rainwater away, rather than collect it or allow it to penetrate the local soils.

According to the National Association of Homebuilders, the average home in the United

EVERY YEAR IN THE UNITED STATES *1.5 million new homes are built, putting a huge strain on natural resources, including timber. Green architects aim to practice sustainability, promoting growth that doesn't deplete natural resources.*

States contains 13,127 feet (4,000 m) of lumber, 13.97 tons (14.19 tonnes) of concrete, 2,085 square feet (194 m²) of flooring material, and 2,427 square feet (225 m²) of roofing material. Multiply this by the 1.5 million new homes built each year in the United States, and the resource implications are astronomical. Then there is the projected global population growth of 1.5 billion people by 2020. Put these three sets of figures together and it's clear that there is an urgent need to design new buildings and developments which will slash consumption of environmental resources.

Green architecture is aiming to tackle the legacy of World War II: vast sprawling suburbs built specifically to provide urgently needed and affordable homes for the young families which were to become the baby boomer generation. Development was segregated by use, with homes in one area, shopping malls in a second, and industry in another, partly because this type of production efficiency had worked so well in the automobile industry. Unfortu-

nately, the results can be seen today in the sprawling, automobile-dominated communities that have been sapping our natural resources.

A ROW OF HOUSES *near Habban in the Yemen (above) is made from mud bricks. Green architects aim to rid buildings of substances that harm the environment, such as CFCs, and replace them with materials from local and sustainable sources.*

BUILDINGS, CITIES AND MARKETS

Green architecture is based around the principle of sustainability—that is, growth which does not deplete natural resouces or harm the environment. A green architect (sometimes called ecological architect) will aim to design buildings which are free of substances that harm the environment in their manufacture, such as CFCs, polyvinyl chlorides (PVCs), and hydrochlorofluorocarbons (HCFCs). Construction materials should come from local sustainable sources, such as plantation timber, straw or earth, or demolished buildings. In addition, buildings should require minimal energy for heating and cooling, and enable the occupants to use water efficiently, then reuse or recycle it.

Green architecture can best achieve its aims when it is done on the level of a whole com-munity rather than individual dwellings. Such communities are created on a more human scale than most cities today. They are designed with the idea of people walking and standing, rather than driving and sitting, and rely on a transport mix of foot, bike, electric car, and public transport. Food is supplied by local agricultural markets to cut down on transport pollution. People take advantage of information technology and work from home rather than drive. Often, such sustainable communities rely on local power generation from renewable energy sources such as solar and wind power.

Achieving this type of change will depend as much on improved design as on advances in new technologies. For example, huge gains in energy conservation will be made by properly

THE GREEN GAMES

Green architecture took a jump forward with the Sydney Olympic Games in September 2000. Local architects submitted a proposal that it should be a "green" games—in other words, that the Olympic Village where the athletes were housed should be built along the principles of green architecture. In the development, waste water is treated on site in artificial wet-lands and buildings are designed to be highly energy efficient. The whole development is built on a remediated, former industrial site. After the Olympics, the athletes' village became a residential suburb.

insulating walls, floors, and roofs. Insulation will be made from renewable sources, such as recycled paper pulp treated to be fireproof. One home recently built in Denmark is so well-insulated that the heat generated by the occupants—two adults, two large dogs, lights, and appliances—meets 50 percent of all heating needs. More buildings will use systems such as heat exchangers and heat pumps to recover and reuse heat generated by the home. In hotter climates closer to the equator, where buildings need to be cooled down, air-conditioners that produce greenhouse gases and rely on CFCs will become a thing of the past. Instead, buildings will be designed to capture natural cross breezes and to take advantage of the stack effect, where hot air rising inside the building generates a cooling draft.

Some green architects will also use high-tech solutions, such as smart

IN THE FUTURE *it is hoped that well-designed communities, when coupled with advances in technologies, will reduce the need to travel by car and will put most destinations within bicycle-riding distance. In so doing, the potential savings in time, energy, environmental damage, and natural resources could be enormous.*

windows that respond to changes in sunlight by darkening or clearing. However, many green architects believe future solutions will be simple and less reliant on technology.

Reinventing towns and cities to be sympathetic with green architecture will require some major changes to the market place. Most developments, for example, are currently designed to save on costs up front, regardless of any environmental impact in the future.

One way of getting around this is to reward companies for reducing emissions in their building design and construction through a carbon credit arrangement. The total emissions for a building would be calculated based on the amount of energy that is used in construction and on the ongoing energy requirements; this would then be compared with a legal limit or standard. If the building exceeded the standard, it would incur a carbon debt; but if it achieved emissions below the standard it would gain credits which could be sold. For the scheme to work, a national standard for greenhouse emissions from the building industry would have to be put in place and then strictly enforced.

SUSTAINABLE COMMUNITIES

Despite these barriers, some experimental examples of green architecture-based communities are being established in the United States. One is the sustainable community of Civano in Tucson, Arizona. This 1,200-acre (485-hectare) community is being built in several stages and is expected to eventually house 2,600 families. The community will combine homes, workplaces, shops, schools, and recreation all within walking distance, to reduce dependence on cars. The design includes both energy and water conservation measures and incorporates the natural environment. For example, local plant species have been rescued from areas destined for roads and buildings, and replanted in green space on the site. The houses have been designed to reduce water consumption and to take advantage of active and passive solar energy. Its designers say Civano's energy and building code will save enough energy to reduce carbon emissions by one billion pounds (454 million kg) over the next two decades.

The aims of green architects have many similarities to another group of architects calling themselves the New Urbanists. New Urbanist developments can be seen in the United States at Seaside, Florida; Kentlands, Maryland; and Laguna West, California. These developments represent a growing trend by Americans in demanding a more pedestrian-friendly community that has not been alienated by cars and freeways. Such new developments typically have a mix of different income-level families, and are carefully designed so that home, work, and recreation are within walking distance of each other.

The Laguna West development in California is an example of what some architects call a pedestrian pocket. First proposed by architects Doug Kelbaugh and Peter Cathorpe, this type of devel-

THE TOWN OF CIVANO in the United States (community center, above) aims to reach a balance between growth, affordability, and achieving a greater integration with the environment, promoting innovation in construction and design.

opment is no more than 100 acres (40 hectares) in size, housing 5,000 people and providing jobs for 3,000, with a density of 50 people per acre. A mix of housing, shopping, community facilities, and employment is linked together, and to major urban centers by light rail. The size of the community is designed to make light rail economically viable and ensure the vitality and human interaction of a traditional small town. As air pollution gets worse in car-dominated cities like Los Angeles, such developments could be models of how in the future we can make towns and cities livable once more.

EXTENSIVE USE OF SOLAR POWER *among sustainable communities will have both a local and a global beneficial effect, providing a source of cheap fuel and electricity, and reducing air pollution.*

Energy

The key question in the energy debate is shifting from "Will we run out of fuel?" to "Whose atmosphere is it, who most wants it cleaned up, and which is the best fuel for the job?"

For Stone Age people, learning to use fire was a most vital developmental step and one of our first encounters with "a technology." By the end of the 20th century, combustion technology had taken us to the Moon, fired the cylinders of billions of cars, and been used to produce more than half the electricity for the switches that dot the walls of modern homes, stores, and offices.

But the scientists of the 21st century are now charged with breaking up this eons-old love affair between people and fire because of relatively new concerns over how the two compete for a common atmosphere. A tantalizing spectrum of emerging energy alternatives could make their job a lot easier.

AS THE 20TH CENTURY CLOSED, *geologists had identified enough coal for another 150 years based on demand at the time. But fears about the environmental cost of such fuels should see the emergence of a new primary fuel this century.*

By burning the fossil remains of life forms from hundreds of millions of years ago, we are pumping massive amounts of carbon from the earth into the atmosphere as carbon dioxide. This traps excessive radiant energy from the sun, which scientists believe causes changing weather patterns and a warming global climate known as the greenhouse effect.

To halt this, scientists want us first to kick two big black habits. Globally each year we burn 4.5 billion tons (4.57 billion tonnes) of coal and about 3.9 billion tons (3.96 billion tonnes) of oil. This is enough black mass to cover the 22-odd square miles (57 km²) of Manhattan to a height of 1,350-feet (410-meters) every three years. And burning a ton (just over a tonne) of fossil fuels puts around 2.5 tons (2.54 tonnes) of carbon dioxide into the atmosphere.

The political shocks of greenhouse are breaking down the acceptability of simply

PROMISING ENERGY *sources of the future include wind turbines like those in California in the United States, pictured left. It is estimated that wind now supplies about one percent of California's electricity. With its lack of green-house gases, uranium (held below) may also regain favor in electricity production.*

burning whatever fuel we can get hold of. But beyond any greenhouse concerns is the indisputable fact that smog from coal and oil kills and hospitalizes tens of thousands of city inhabitants every day. Nitrous oxide and sulfur dioxide are also major emissions that continue to dangerously acidify rain in many parts of the world, and the more sulfur-rich fuels are being banned or economically penalized. All in all, a rather grim challenge has been set for a century of energy innovation and diversity.

But a cavalry of scientific help is on its way. Energy scientists are changing the entire complexion of power generation using renewable resources like wind and solar energy, and non-combusting technologies like fuel cell reactions.

See-through films that act as solar panels, applied directly over the windows of homes and office blocks, could be the next renewable energy breakthrough, as could super-efficient windmills with 260-foot (80-m) blades. Fuel cells acting in reverse could then produce hydrogen to store the renewable energy. But most importantly, the self-defeating barriers to alternative energy markets, such as institutional lending practices and lack of consumer finance, are being recognized. With this recognition, renewable energy systems could finally overhaul the unrelenting 1.6 percent growth in fossil fuel consumption per annum.

Even the production of electricity by the fission of uranium, traditionally a politically tainted technology, is regaining favor for its lack of greenhouse emissions, so that the ranks of 435 operating nuclear power plants may swell.

The 21st century will be the Colosseum in which energy technologies and the commercial interests behind them slug it out and diversify our energy diet. One of the biggest growth areas is the revolutionary fuel cell technology, with the first cars, mobile phones, and domestic power units now being marketed.

FUEL CELLS—GIVING UP SMOKING

Despite its high-tech image, the fuel cell was first discovered in 1839 by William Robert Grove when he used a set of four fuel cells to drive an electrolytic cell. In a jackknife of history, Grove's gas battery had no obvious use at the time as the light bulb and electric motor had not yet been invented.

POWERING AN UN-GLOBAL VILLAGE

In the poorest developing nations, there is still a clamoring for essentially 19th-century energy technology. How will such countries afford to make the shift to fuel cell technology? Paradoxically, while the poorer nations use the basic methods to burn fuels, causing heavy pollution and greenhouse penalties, a massive 84 percent of energy is consumed by the wealthiest 20 percent of the global population.

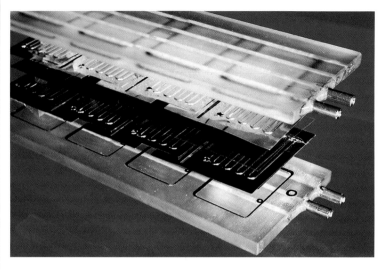

AT THE HEAD OF *new energy technology is a layered structure membrane fuel cell, which converts hydrogen to electricity. Developed at the Fraunhofer Institute for Solar Energy Systems in Freiburg, Germany, the fuel cell was designed for the prototype of a hydrogen-powered notebook computer. The fuel cell (left) occupies minimal space yet still provides the output voltage required for a portable computer.*

But now its time has come. Fuel cells of today incorporate Grove's three essential components—the cathode for reducing air or oxygen as the positive pole of the cell, an anode where the fuel is oxidized as the negative pole, and an electrolyte for the selective transport of ions between the two. The reaction is highly efficient in producing electricity. It is fully contained, and the principal limitation of the device's lifetime is the slow corrosion of the electrodes.

Basements and garages will increasingly be the site of domestic power generation in the 21st century, with a 3 cubic foot (1 m³) fuel cell quietly, safely, and efficiently reacting a fuel—perhaps methanol, propane, or natural gas—as the source of the key ingredient, hydrogen. Adding to the efficiency of fuel cells is the slashing of power line transmission losses.

The Tokyo Electric Power Company has been demonstrating a phosphoric acid fuel cell since 1991, with an impressive efficiency of 43.6 percent. At 11 MW capacity, it produces enough energy for a small suburb. But eventually the fuel cell in the family car may out-supply such centralized units. By plugging the car into the grid during the 96 percent of the time when it's parked, and selling excess power generated by the fuel cell back to the grid, motorists could earn back as much as half the cost of owning a car.

But the switch to fuel cells and a "hydrogen economy" must make use of economic and philosophical

ROBERT HOCKADAY *of Manhattan Scientifics holds a sheet of fuel cells that are being developed in an effort to commercialize a long-life Micro-Fuel power cell power system for portable electronics like telephones.*

strengths to overhaul both the coal and nuclear plants. Using a slightly different technology, an ambitious plan by Norsk Hydro hopes to bring a hydrogen-burning power plant on stream for $1.3 billion. Located in coastal Norway, it will produce 120 MW and reinject carbon-rich fluid wastes back into the sedimentary formations they came from.

WAFER POWER FOR YOUR PHONE

At the microscopic end, Robert Hockaday, a former Los Alamos physicist and now principal of Manhattan Scientifics Inc. in the United States, has developed a fuel cell battery that is produced in mile-long (1.5-km-long) sheets.

Manhattan Scientifics is now working to extend the life of these microscopically layered wafer cells from 600 hours to 3,000 hours. A hungry army of portable consumer electronics such as computers, organizers, and mobile phones, awaits. The mobile phone cell will run on a combination of water and methanol from a tiny squeeze bottle.

ENERGY RIDDLES AND THE ICE THAT BURNS

Deep in the dark abysses of the world's oceans, massive slabs of a strange, white, icy sediment have been discovered. With as much contained energy as all other fossil fuels put together, methane hydrates may be the cleaner fuel we have been looking for, or a major contributor to Earth's greenhouse effect. They are, at least, a new and fantastic geological surprise.

Methane hydrate looks a lot like water ice. It is a crystalline solid consisting of gas molecules—usually methane—caged by water molecules. About 15 percent of the weight of methane hydrate is methane. Dropped on the deck of a ship, it jumps around like frying bacon as it releases 160 times its volume in gas.

The United States Geological Survey (USGS) has estimated that, worldwide, the amount of carbon found in gas hydrates totals twice the amount of carbon to be found in any other fossil fuel, and that methane in hydrates amounts to approximately 3,000 times the volume of methane in the atmosphere.

Methane ice is laid down in icy depths of 1,500 feet (460 m) or more. As great tongues of 35°F (1.5°C) water continually circulate from the ice caps, they cool the under-stories of the oceans and freeze methane hydrate formed from organic debris that has been washed from the continents. It then drifts slowly to the bottom of the ocean, settling in snow-like blankets.

The discovery of methane hydrate ice was way back in 1981 aboard the *Glomar Challenger* when a yard-long (0.9-m-long) core was recovered. But it has been the recent discovery of astonishing volumes of methane hydrate that has launched it from curio status to the subject of massive research efforts.

Major gas importer Japan is currently the leader of those considering the "ice that burns" as a future fuel, perhaps for fuel cells and is working on a demonstration project of hydrate extraction in the Nankai Trough off the east coast of Japan's main island.

Whether it becomes our next fuel or not, methane hydrate may already be playing a key role in the make-up of the atmosphere. The USGS believes the breakdown of methane by radiation in the upper atmosphere could be the overwhelming cause of variations in carbon dioxide levels. Methane is a gas 10–20 times more effective than carbon dioxide in swinging Earth's radiation ledger and its resulting greenhouse effect. The drilling of sediments off the east coast of North America in 2000 produced evidence in beds dating back 55 million years that a huge eruption of methane from the seafloor triggered a massive global warming event. Scientists believe this event allowed the mammals to migrate more widely and begin their ascendancy in the Age of the Mammal.

Submarine landslides, geological upheaval of sediments, seafloor vulcanism and changes in ocean temperature can cause methane hydrates to thaw and to release massive gaseous eruptions. In fact, global warming could itself create a chain reaction that releases more methane from the oceans.

On the other hand, if methane from these super-abundant hydrates was used widely as a fuel, this might help reduce greenhouse emissions. When methane combusts it offers lower carbon dioxide emissions than any of the other hydrocarbons now in global use.

One thing is clear. Any radical shift made in the energy industry will depend on economic efficiency and the degree of importance placed on degradation of the environment.

METHANE FROZEN *as a hydrate in seafloor sediments potentially represents a massive alternative energy source for the future and is being considered as such. It may also be an unpredictable greenhouse gas.*

The Future of Sport

The late-20th-century doping crises will fade into insignificance as the world's sporting associations face a torrent of ethical issues linked to 21st-century technological innovations.

The best coaches will always have a knack for recognizing and nurturing the sort of raw talent that separates sporting champions from also-rans. But as the 21st century progresses, identifying prospective sporting superstars and helping them achieve their full potential will owe more to science and technology and less to gut instinct, experienced hunches, and educated guesswork.

TECHNOLOGICAL TEAM-WORK

Many elite athletes already engage teams of high-tech specialists to help them maximize the effects of training and perform at their optimum during competitions. Some won't begin a pre-match psyche-up without a psychologist, swallow a morsel of food that hasn't been sanctioned by a nutritionist, or set out on a new training regime before receiving advice from a biomechanist or kinesiologist armed with video and computer equipment.

And then there are exercise physiologists employed to track metabolic changes in blood and muscles during training and competition, biomechanical engineers designing and building improved sporting equipment, and sports doctors applying treatments that substantially accelerate healing and reduce injury time.

In the near future, geneticists are likely to become routinely involved with the sporting elite. As it's widely accepted that most sporting greats are born with an innate competitive edge, scientists are trying to map the genes that underlie athletic prowess. One of the earliest breakthroughs came in the late 1990s when researchers from Britain and the United States identified a gene known as the Angiotensin Converting Enzyme (ACE) gene which may play an important role in athletic endurance.

Experts believe parents will eventually be able to test the DNA of their children to dis-

NO MATTER HOW SPORT *develops throughout this century, raw talent will always be the key ingredient for success. Genetic testing could one day enable us to identify that talent at a very early age, long before it has manifested itself.*

cern whether they're likely to grow up to excel in the sporting arena. The ability to identify specific athletic pursuits to which people are best suited could follow. And later in the century it could even become possible to genetically engineer offspring with advanced sporting potential.

PLAYING RECORDS WITH SUPERMEN

What will all this progress mean? High-jumpers able to leap tall buildings and runners faster than speeding bullets? No one understands exactly how all the current and emerging scientific support is going to affect the future of athletic endeavor, nor does anyone know how much humans can continue to improve (or be improved) athletically.

Many experts believe there must be a limit to the physical capabilities of our species and that sporting record books can't continue to be rewritten indefinitely. But then, it was widely assumed little more than six decades ago that no one would ever run a sub 4-minute mile. That barrier was officially broken for the first time at Oxford in 1954 by Britain's Roger Bannister. Now 3 minutes 30 seconds is the miler's next big target and that record, according to a mathematical model produced by researchers at Canada's McGill University during the late 1980s, could be expected to fall sometime around 2040. A few years before that, the model predicts, a woman will run a sub-four minute mile for the first time.

As records continue to topple early in the 21st century, heated debates about regulating and even banning a range of performance-enhancing scientific and technological interventions are anticipated. The drugs that sparked scandals in so many sports during the late 20th century will seem benign in comparison to future possibilities. How, for example, will sporting purists feel about a muscle-building "strength virus" currently under development that builds bulk and power without exercise?

The virus is likely to be delivered into the body by hypodermic syringe. Some of the

FASTER, STRONGER, BETTER—*records offer sports stars a kind of immortality. But how much assistance from science will nature need in the future to keep records tumbling?*

more controversial advances, however, will probably involve surgical procedures or medical interventions that modify the body in some way. Already some baseball and NFL players in the United States have had their vision improved by a special form of laser surgery that modifies the corneas.

Advances in transplant technology, prosthetics, and eventually in cybernetics (bionic implants) could ultimately make it possible to improve athletic performance with the help of new body parts. Such advances would certainly blur the line between able-bodied sports stars and those who currently fall into the category of "disabled" athlete.

If sport does develop along such a highly-interventionist path, *the* big question for the future may well be: Is this sport at all?

Electronic Democracy

Today you can do your banking, buy a book, and even go to school on the Internet. But will computers and the Internet really make the world a fairer, more democratic place in which to live?

Artificially intelligent beings will have the right to vote in political elections; governments will base their decisions exclusively on electronic opinion polling of their citizens; and each of us will vote regularly on the issues of the day, using our mobile phones. That's one vision of a totally wired democracy, and it's closer than you may think.

The Greeks first defined *demokratia* (*demos* meaning "people" and *kratos* meaning "rule") as "the rule of the people." For most of us, however, exercising our democratic rights means turning up to the polling booths once every few years, and then leaving the rest to a handful of elected representatives.

In the 21st century the way we practice democracy is being radically reinvented. With the help of remarkable new technologies, the "voice of the people" has the potential to be louder and more influential than ever before. Through the click of a computer mouse, new digital communities are forming every day via the Internet, and mobilizing themselves to shape political agendas.

AUTOMATED POLITICS

As we move toward a paperless society, it seems inevitable that elections will become electronic too. Some believe that a fully computerized voting system will ensure that vote counting is quicker and less prone to human error and intervention. In fact, the technology is already in place to allow you to vote from the comfort of your own home, local library, or workplace, using the Internet or even your telephone.

In some US states computer polling booths are already in use, and during the 2000 United States presidential elections a few hundred overseas military personnel were able to cast absentee ballots over the Internet. Internet voting will happen on a large scale eventually, but some big problems will need to be ironed out first. Major elections are an attractive target for hackers, who could vote more than once over the Internet or even "crash" the process. Protecting voter privacy and authenticating voter identity will remain one of the trickiest challenges for online voting.

THE VOICE OF THE PEOPLE

Ideally, a true democracy is one in which an elected government takes into account the opinions of its citizens. British futurologist Ian Pearson says that, through electronic networks, it is already possible to construct a massive database that records the preferences of every individual in the society. He predicts that one day each of us may possess an electronic

"shadow," as he describes it, that defines our preferences on various issues.

"Instead of electing a representative every four or five years, the shadow would be available 24 hours a day to allow the electorate to make their opinions known on every issue. Voters could then modify their shadow at will, or be lazy and just select the default settings offered by political parties," Pearson says. "By inspecting all the shadows, and running any algorithms, a government would then know what the population wants—effectively a referendum on every issue without everyone having to go out and vote."

This vision for a direct democracy conjures up some scary scenarios. As Pearson puts it, "if we ever have a direct say in everything, we'll probably make some daft decisions, such as voting for free chocolate and no taxes—then we'll all die of heart attacks while our economy grinds to a halt."

DEMOCRACY WITHOUT ACCESS?

Whatever exciting possibilities lie before us, the computer cannot become a tool for democracy unless we all have access to the technology. Half of the world's population has never made or received a telephone call, let alone used a computer. A computer costs the average Bangladeshi more than eight years' wages, compared with one month's income for the average American.

Without equity of access between all people—men and women, rich and poor, educated and illiterate, young and old—a truly democratic electronic democracy will remain little more than a fanciful dream.

PROTESTERS GO CYBER

Critical Mass is a pedal-powered group of protesters who meet each month in cities all over the world, disrupting traffic wherever they go. Behind the scenes is a busy community of car-hating web aficionados. Using websites and e-mail lists, these cyber activists publish photos, audio recordings, and news of Critical Mass events everywhere.

In 1999, thousands of people took to the streets of Seattle in the United States, to protest against the activities of the World Trade Organization. They took to the Internet too, and out of their vocal campaign was born the Independent Media Center (above). This international project has developed its own exciting new technology that anyone can use to publish their own news stories, photos, and video and audio clips on the growing number of Indy Media websites. The organization aims to provide a democratic and grassroots media outlet by encouraging people all over the world to report on events in their own communities.

A LONG LINE *of people waits patiently to vote in Nelson Mandela's native village of Transkei, South Africa. The nation's black citizens were not entitled to vote until 1994.*

Tackling Crime

*Law enforcement agencies are closing in on traditional crimes
but cyberspace threatens to develop into a lawless frontier
overrun by crooks with computer savvy.*

I t doesn't take a lot to convict
a killer these days. Saliva on
an envelope flap or a tooth-
pick can help to send a criminal
down. So can a speck of blood at
an otherwise pristine crime scene.
Such biological remnants can be
crucial evidence in modern criminal investigations because they
contain DNA, and other than
identical siblings, no one has
exactly the same DNA, making
it a practically infallible mark of
a person's identity.

NOWHERE TO HIDE

Technology used to analyze and
compare DNA—known as DNA
"fingerprinting"—was arguably
the 20th century's single most significant forensic science breakthrough. Since its development in
1985 it has been applied globally
in law enforcement, using only
blood and semen at first but now
also dead tissue, such as hair.

Currently, DNA fingerprinting
is used largely to ascertain the
presence or absence of particular
suspects at crime scenes, but it's
also likely to become a routine investigative
tool. Researchers are developing mobile hand-
held analyzers to assess DNA at crime scenes
and rapidly check it against remote databases
of previously convicted felons. Ultimately,
investigators may even be able to deduce the
appearance of unidentified suspects from
crime-scene DNA.

Many other recent technological innovations challenge the abilities of crooks to cover
their tracks. Chemical analysis techniques have
become so sensitive in recent years that cus-
toms officers can detect residues on passports
from bomb making or from drug smuggling.
Infrared microscopes can reveal shoe polish
scuffs left on clothing by an attacker's kicks.

MOST DNA EVIDENCE *currently ends up in a laboratory for
analysis. In the not too distant future, however, it's likely that
forensics experts will be able to analyze DNA evidence before
leaving a crime scene by using portable, hand-held equipment.*

Similar light technologies have been developed
to expose other seemingly invisible clues, including signs of document tampering, latent
fingerprints, or concealed weapons.

LAW ENFORCEMENT GADGETRY

As long as criminals use guns, so will crime
fighters. But the holsters of future cops could
carry non-lethal varieties. Research into human
behavior modification through sound waves is
among many promising alternatives. Acoustic

barriers and guns that stun or temporarily disable by vibrating internal organs could soon be used to control rioting crowds.

High-tech gadgetry is also making it easier to keep tabs on convicted criminals. Electronic identification and tracking devices that require surgical implantation have been developed, but how extensively such devices might be used is impossible to predict. Tagging with electronic anklets and bracelets after release from prison has, however, already begun in some countries.

Future criminals will need to be suspicious of every fly on the wall because one may well be a tiny mobile surveillance camera. And they will also need to be wary of their own computers if they come with sound cards and video cameras installed. Such equipment can be covertly accessed and switched on by security operatives with the right technological know-how from anywhere in the world.

COMPUTERS: ALLIES AND ENEMIES

Crime fighting is also being transformed by the massive capabilities of computer and information technology. Front-line police can now rapidly access information in huge remote databases using mobile equipment small enough to be hand-held or installed in a car.

Detectives are discarding wall maps dotted with colored pins and string. Instead they're plotting and predicting crime trends in a fraction of the time with computerized mapping, and are reconstructing crimes using virtual reality technology.

FROM BEAT COPS *checking license details of a car seen running a red light to homicide detectives mapping the modus operandi of a serial killer, police around the world are already finding computers indispensable tools in the fight against crime.*

Although computers are improving policing they're also creating intractable havens for illegal activity. Theft, fraud, embezzlement, extortion, stalking—they're already happening on the Internet in a big way and growing at a phenomenal rate. Crimes with new names, such as electronic vandalism, hacktivism, and cyber-terrorism are also rising rapidly. Even organized crime syndicates are coordinating operations via the Internet and using it for money laundering.

It will take global laws and international teams of technologically skilled cybercops, technospies, and computer forensics gurus to bring Internet-related crime under control.

IDENTITY THEFT AND BIOMETRICS

Once dubbed the "crime of the 90s," identity theft is forecast to become the "scourge of the 21st century." Experts believe that, in the United States alone, there may be more than 500,000 new cases annually.

Perpetrators use other people's identification details to access savings or credit accounts and commit similar frauds. Stolen credit cards, drivers' licenses and similar documents used to be prime targets. But the Internet has expanded opportunities and fuelled growth of this crime. One answer will be new "biometric" methods of personal identification, which recognize unique physical attributes. Fingerprint, retinal, and iris scanners could become commonplace. So might systems that recognize voices, handprints, and hand vein geometry.

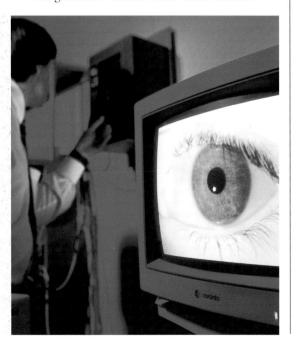

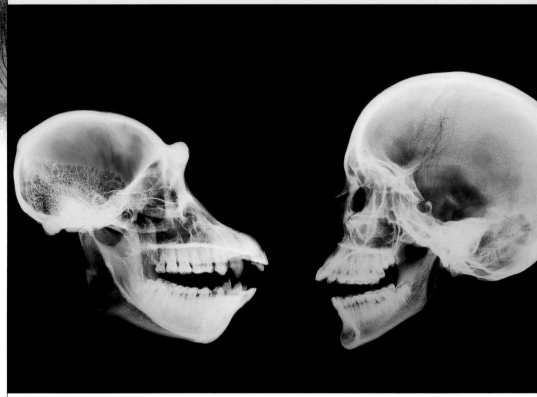

Understanding the Past

The ancient world holds many mysteries, the answers

to which continue to elude us. Now new technologies offer

powerful tools to explore and reveal that world.

As the technologies of the future advance, so the sciences of the past—archaeology and paleontology—will continue to dig up startling new discoveries. Already, biochemistry and genetics are helping to resolve the deeply contested debate about where we as a species came from. Genes change over time and, by measuring genetic changes in different peoples of the world, relatively new technologies have demonstrated that 98.4 percent of our DNA is identical to that of a chimpanzee. The rate of genetic change indicates that chimpanzees and humans shared a common ancestor somewhere between 5 and 7 million years ago. Many subsequent fossil finds support this theory.

Genetics has also confirmed that humans evolved in Africa within the last half million years and subsequently spread out across the world. Other studies show the complex inter-relationships and movements of modern Europeans, and demonstrate that the Polynesians of the Pacific originated in Taiwan. The story

of human evolution will be told in ever-finer detail as these technologies develop.

BRINGING BACK THE DEAD

One extraordinary possibility is offered by genetics; namely, bringing extinct animals back to life by cloning their DNA. While dinosaurs may have been dead too long, we could re-create more-recently extinct animals. The key is to find intact DNA. Recovering this from an organism that has been dead for more than a few thousand years is unlikely because DNA is

THE WOOLLY

mammoth (right)
became extinct
some 10,000
years ago. Cloning
its DNA may give it new life,
but the species will also need
the right environment to survive.

NEW TECHNOLOGIES *have proven that we share a high percentage of our DNA in common with certain primates (left). It is hoped that radar images of Angkor Wat in Cambodia (right), may unearth temples previously hidden from view.*

delicate and breaks down over time. But there is a possibility of finding intact DNA of organisms that became extinct more recently.

Japanese researchers are currently trying to extract DNA from frozen woolly mammoths, with the hope of implanting it in the nucleus of a living elephant's egg. From this, a clone of the original animal could develop. Similar efforts are being made to revive the thylacine, a wolf-like marsupial from Australia, extinct since 1936. If successful, this technique could be applied to several other animals, including the passenger pigeon and the dodo.

BURIED TREASURE

The buried ruins of ancient buildings and towns leave slight changes—in moisture content, height, and nutrient levels, for example—on the Earth's surface. At ground level these changes can be imperceptible, but from above they leave tell-tale patterns, which are possible to detect by satellites and other mapping technologies. Satellite images cover huge areas in extreme detail. Archaeological sites, however, tend to be small. So computer programs are being developed to search the actual satellite images for hidden archaeological treasures.

Satellite images can also be used to map existing sites across continents to give a picture of the large-scale activities of lost civilizations and the interactions between their temples, cities, villages, and agricultural regions.

THE PAST IN CYBERSPACE

Computers and digital technology first developed to make advanced computer games are now being used to recreate the ancient past—virtually. Ruins and rubble can be pieced together using a computer, and missing pieces added. The finished virtual building can then be rendered using a 3-D animation program. Australian researchers have recreated a Roman theater in this way. It enables visitors to "walk" the long-lost corridors and "sit" in the once-ruined auditorium. Soon, they hope to introduce virtual actors, allowing us the opportunity to watch a Roman play as and where it was performed approximately 2,000 years ago.

We can also now look into the face of a person who has been dead for thousands of

VISITORS TO *the forum ruins in Pompeii, Italy, can only imagine what the city was like in 79BC. It is hoped that virtual reality will make it possible to take a "journey" into the past.*

years, thanks to the techniques developed to reconstruct the skulls and faces of murder victims or missing persons. Ancient skull material is CAT-scanned and digitally manipulated to give a virtual reconstruction, then a solid copy is produced using a technique called 3-D resin lithography. Modeling clay builds up muscles, fat, and skin on the resin skull, thus reconstructing the ancient face.

With greater computing power and more accurate mapping of ancient sites in the future, the recreation of entire virtual cities could be achievable, giving us the experience of, and insight into, life in ancient societies. As for the recreation of the original inhabitants of those societies, virtual skulls could be dressed in skin and muscles generated by computer, then animated to produce walking, talking reconstructions. Animals could be reconstructed in the same way. One day, it seems, *Jurassic Park* may be a reality—although only within the virtual world of very powerful computers.

CHAPTER THREE
FRONTIERS *of*
TECHNOLOGY

Any sufficiently advanced technology
is indistinguishable from magic.

The Lost Worlds of 2001,
SIR ARTHUR C. CLARKE (b. 1917), British writer and scientist

The Magic of New Materials

The discovery of a wealth of new materials is helping us to create everything from artificial hearts to automobiles and clothes that can change color with the flick of a switch.

The last decade has seen an explosion in remarkable new materials science. Certainly we are well equipped when it comes to designing new materials: There are almost a hundred chemical elements that we can combine in all kinds of novel ways. We are learning to turn them into ultra-hard coatings that can work inside the body, to make light-bulbs that never die, and to create vehicles that change color at the push of a button.

However, one element in particular has proved itself the most versatile building block of all—carbon. This first became clear after chemists Harry Kroto, Richard Smalley, and Robert Curl stumbled across a brand new form of carbon—C60—which contains 60 atoms of carbon. They named it Buckminsterfullerene, after Buckminster Fuller, the architect of the geodesic dome that the C60 molecule resembles. The discovery of this form of carbon

BEFORE THE DISCOVERY *of the buckminsterfullerene, carbon was known to take two main forms: diamond (below) and graphite, plus irregular forms known as amorphous carbons. It is now responsible for the smallest wires ever made—buckytubes.*

came as a complete surprise—all the more so because the chemistry of carbon is one of the most studied of all the elements.

Yet buckminsterfullerene turns out to be just the tip of the carbon-based iceberg. Its discovery has been followed by a stream of other bizarre new forms of carbon. There are nano-tubes of pure carbon whose walls are curled-up graphite sheets. These tubes are tiny, with a diameter 10,000 times smaller than a human hair and they could one day be used as tiny filters or electrical components. Then came double-walled nanotubes and Y-shaped tubes, and, most recently, structures that resemble pea-pods—nanotubes that hold several C60 molecules within them.

We have now learned that we can make a huge variety of structures from simple carbon fibers by using the same principles as those used for centuries by Japanese basket weavers to create artistic shapes. With these recipes, it is possible to reproduce practically any object in pure carbon, with the result that they are lightweight, ultra-tough, and have the ability to withstand high temperatures.

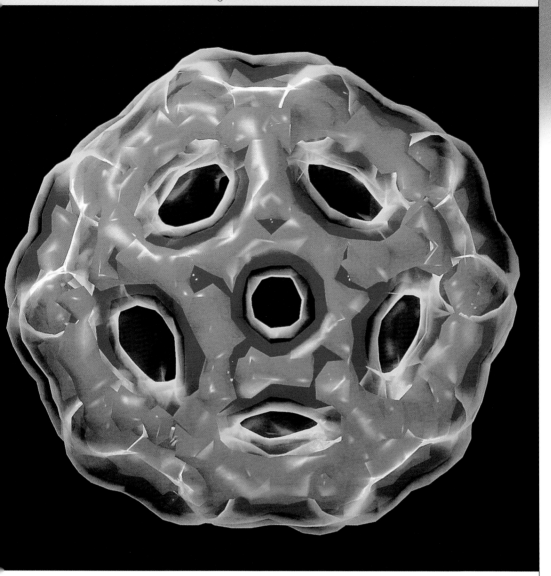

Some of these objects are already being put to good use: For example, thin nanotubes with pointed ends have been found to emit electrons. These could be useful for building thin, lightweight displays for television screens and computer monitors. In the future we should even be able to build tiny working machines, complete with gears and axles, from nothing more than carbon nanotubes.

Carbon fibers are already being used to make parts for jet engines and space rockets.

NEW BODY PARTS

Carbon has also provided us with yet another remarkable material—this time in the form of an artificial diamond that is glassy or amorphous, rather than crystalline like natural diamond. Amorphous diamond is made by heating carbon in an electric arc until it is so hot it forms a plasma. Atoms of carbon inside the plasma then condense on any surface in contact with it to form an ultra-tough coating. In fact

THE DISCOVERY *of the all-carbon soccer-ball molecule known as buckminsterfullerene in 1996 (as a computer-generated image above and below under a microscope) opened an almost-closed book on the structure of carbon. Many chemists see buckyballs as important building blocks for the future.*

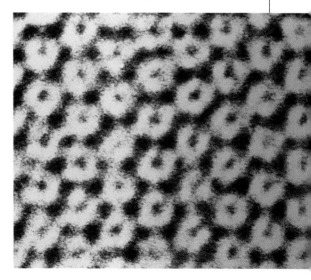

this material is almost as hard as natural diamond. But since amorphous diamond is relatively simple to make, it can be applied as a coating on all kinds of surfaces. Applications for this may include things like long-life razor blades, and protective coatings on computer hard drives and high-speed drills. Since it also conducts heat, artificial diamond could be used as a heat-dissipative coating on computer chips.

Carbon has also been found to have excellent compatibility with the human body. Special carbon coatings are under development to provide wear protection and blood compatibility on the interior surfaces of a blood pump that will eventually be used as an artificial heart. Artificial hearts may, in time, take over from heart transplants for those with chronic heart failure as the demand for transplantable organs far exceeds supply.

EVERLASTING LIGHT

Carbon also plays a vital role in one of our most versatile materials. It readily forms long chain molecules with other elements, especially oxygen, nitrogen, and hydrogen to form long chain molecules called polymers. Polymers such as polyurethane or Kevlar have the combined advantages of lightness and strength (as a result of their long chain structure), and can

THE ARTIFICIAL DIAMOND, *or amorphous diamond, works as an extremely hard coating that is almost as hard as natural diamond but relatively simple to make and certainly much more cost-effective. It could be used as a protective coating on many things, including perhaps long-life razor blades.*

be easily formed into a variety of shapes. The surface properties of some polymers can now be modified by shooting ions into them to give the polymers a hardness and scratch resistance similar to that of a metal.

Some polymers such as polyphenylene seem to behave like semiconductors—under certain conditions they can conduct electricity. In a revolutionary development, derivatives of this type of polymer can be made to emit light when they carry a current. And by changing the chemical structure of the polymer, chemists can even tune what color light they give out.

These discoveries indicate we should soon be able to create large light sources for room lighting and display boards. Prototype displays have been created in labs in Cambridge, England, and in California, United States, and they could soon find their way into mobile phones and palm-top computers.

But, as lighting in particular, these materials offer some remarkable advantages. Firstly, they give out little or no heat, so they can be much more energy-efficient. Also, unlike lightbulbs, they can last for many hundreds of thousands of hours before they fail. Eventually, you might coat the ceiling of a room with tiles made from these light-emitting polymers to provide a daylight illumination effect. And you'd never need to change a lightbulb again since these polymers should remain active for a lifetime.

Light-emitting polymers can also be made into flexible sheets that can be rolled up. This offers the possibility that we may one day carry large "smart" screens with us wherever we go. Fitted with a wireless receiver, these screens could display the daily newspaper for us on our way to work, and would take the place of our desktop computers once we arrive at the office.

PERMANENT BUTTERFLY COLORS

Some of the most remarkable materials have already been developed by nature during the course of evolution, over millions of years. We are only just beginning to discover their potential. Take, for example, the brilliant metallic colors found on the wings of butterflies like the electric blue Morpho from Central America, or the greens of the Papilios from South-east Asia.

These species of butterfly have wings that are coated in special scales. Each scale has tiny structures that resemble the cross-section through a Christmas tree. Each "branch" of these trees reflects light in a particular way so that some colors pass through while others

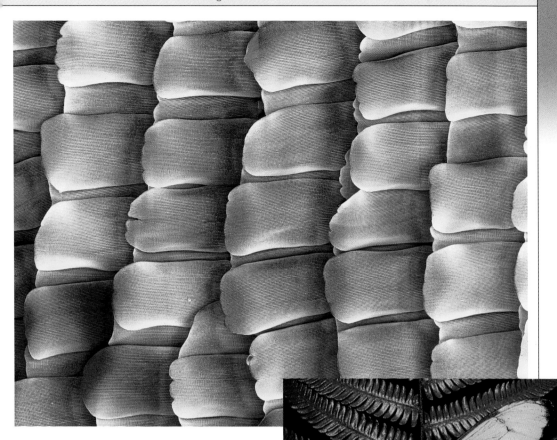

SOME REMARKABLE MATERIALS *can be found in nature. Butterfly scales (seen under a scanning electron micrograph, above) have a metallic appearance due to structures within them. They can produce brilliant colors that do not fade (right) and by studying the structure of a butterfly's wings researchers hope to make materials that could alter their colors.*

are reflected from the wing and are made to appear brighter. Such structures are relatively new to science, and have been named "photonic materials" because of the effects they produce with light.

Butterfly scales have a lot to teach us. They can produce brilliant colors without the need for dyes. Eventually this trick could be useful for making colored images that never fade—unlike old color photographs, for example, which change with time. They may offer wonderful new ways to color everyday objects, too. Car manufacturers are already developing paints based on photonic materials that produce a unique finish which changes color depending on the angle you view it from.

Engineers are also studying butterfly scales with an eye toward making credit cards and bank notes more resistant to forgers. If, for instance, they can develop markings based on these tiny, intricate Christmas trees and turn them into security devices on credit cards, their complex structure and bright colors would be almost impossible to counterfeit. Plastic bank-

notes, too, which have been pioneered in Australia, could be armed with photonic security markings based on the butterfly wing.

But these types of natural materials have even more remarkable powers to offer. By changing the separation between the branches of the Christmas tree, scientists have realized they can tune the colors reflected by the structure. And they have already discovered one beetle that does this in nature, always ready to alter its patterns of camouflage to blend in with the environment. This trick points to all kinds of new applications for these materials. If you can control the separation of the branches, with pressurized air, or liquid for instance, you could make materials that change color at the flick of a switch. Materials in the 21st century may not only be as tough as diamonds, they could alter their colors like a chameleon.

Biomimetics

Nature is a master engineer, and we can learn a great deal

from studying the lessons found in its book of design plans.

Almost every conceivable engineering problem that we face today has been faced by nature in the past, with its powerful tool for solving problems—evolution. We have been borrowing nature's ideas throughout history, but modern science and technology allow us to look even more closely at nature. And the deeper we look, the more there is to find. Nature has even found solutions to problems we are yet to encounter. And the whole process of studying nature to provide technological breakthroughs has now been forged into the science of biomimetics.

Generally, biomimetics (meaning "to mimic life") involves observing features in nature that we can't duplicate yet. Perhaps through studying how natural structures and processes work, we will find new ways of solving old problems.

Biomimetics holds the promise of scientific breakthroughs and imaginative new technologies. We may create fabrics that adjust their insulating properties according to their environment, courtesy of a pine cone's opening and closing mechanism. Oil-based plastics may be replaced by expanded starches (which would be biodegradable and not use dwindling oil reserves), from studying how plant starches work. Mechanical tools could be developed that convert chemical energy into mechanical force, the same way a worm uses hydraulics.

MAKING STEEL

To be "as hard as steel" is not always hard enough. There are many cutting and drilling situations where materials harder than steel are needed, but such materials are currently difficult or impossible to manufacture. Chitons may hold the answer. These mollusks live attached to rocks and they eat the top layer of that rock to obtain microscopic algae. For this they use their teeth, which are coated with magnetite, an iron oxide mineral that is harder than our hardest steel.

We can make poor-quality magnetite in the laboratory, but only under pressures and temperatures that would kill a living organism. But chitons somehow make a very pure form of magnetite without using such high pressures or temperatures, probably as a result of a complex chemical process. Scientists are examining chi-

tons and related mollusks in the hope of manu-facturing magnetite in the same way, so that it can be used to toughen drill bits or coat blades.

STRUCTURAL COLOR

Most colors are produced by pigments, mole-cules that generate color but break down over time, so that colors fade. But nature also has a spectrum of colors, called structural colors, that don't use pigments. They tend to be iridescent and particularly vivid, and may be used by small marine creatures, insects, and birds, because they are easier to make than pigment colors.

British and Australian researchers are just beginning to list the many different types of structural color and how they work. Some sim-ple structural colors can be manufactured with existing technologies, like the rainbow of color on a CD. But many rely on complex micro-scopic geometries, which we currently cannot reproduce. When we can, structural colors may not only give us color photographs that never fade, but their light-handling properties could have several applications in photonics and other communications technologies. Some structural colors absorb all light, giving absolute black, which could be used to coat stealth aircraft, or collect light in night-vision goggles.

BONE BUILDING

Bones are one of nature's ways of handling loads and stresses. Irregularly shaped to be spe-cifically tailored to their task, they are made of

THE CRAB *was the inspiration for Ariel (above), a machine designed to "scuttle" through surf to search for mines on the ocean floor. Nature inspires other advances, too. Colors without pigments, like those in iridescent peacock feathers (opposite) are being researched to create colors that won't fade.*

a composite material, a rigid but brittle frame-work of calcium phosphate interleaved with a flexible protein. This combination of a single, flexible, and adaptable material makes bones exceptionally efficient structures.

The interconnected components in conven-tional structures interact with each other in complex and often unpredictable ways, but this is eliminated in a bone-based architecture because an entire structure can be designed as a single element. The simplicity of building with a single element and the composite nature of the building material offer design solutions for buildings, bridges, and arches that conventional construction materials simply cannot match. Researchers are beginning to learn more about how to recreate these natural features, but bone building remains an elegant idea for the future.

The challenge, then, is to mimic the devices we find in nature and apply them to our own needs. But we must also be ready to learn about the efficient technologies used in nature that do not destroy the environment. Manufacturing techniques that create wood from soil and sun-light, or silken threads stronger than steel made at less than body temperature are just two of a list of endless examples that we can learn from.

Nanotechnology

Imagine tiny machines engineered to microscopic precision,

with each component made up of just a few atoms.

While such miniaturization may not sound like a great revolution, its implications—and applications—will be profound. Known as nanotechnology, this emerging field is more than simply an engineering feat. It could open the door to dramatic improvements in antiaging treatments, medicines, and supersmart and superstrong materials. It may also allow goods to be manufactured in new ways: on submicroscopic production lines, manned by tiny robots. And according to its proponents, this technology is only decades away.

In some ways, nanotechnology is an extension of the technology used to build computer chips. The name comes from the size of the smallest components engineers are able to build with. Today's computer microchip circuitry is only a few micrometers (millionths of a meter) across, so the state of the art is "microtechnology." Nanotechnology will have truly arrived when manufacturers can assemble parts that are a thousand times smaller than this—components just a few nanometers (or billionths of a meter) wide. Once such a level

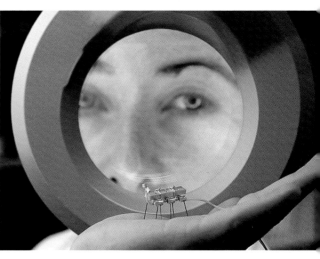

TODAY'S TINIEST ROBOTS *will be relative giants next to the germ-sized machines of the future, say advocates of nanotechnology. A nanomachine, for instance, would be far too small to be even visible on the palm of someone's hand.*

is reached, manufacturers will, for the first time, be able to make computer components in which each atom is purposefully and precisely placed where it is required, saving on space and creating greater efficiency. And components will consist of only a few atoms each.

Imagine the minuscule machinery that will be made possible, and how powerful computers far smaller than a speck of dust might be. Such computers could be connected to equally tiny robots, known as "nanobots," creating submicroscopic machines that could perform seemingly miraculous medical tasks.

For example, such nanobots could be programmed to seek out and kill cancerous cells. An army of them could be injected into cancer sufferers, to patrol their bodies day and night, and forever keep them free of cancer. They might be programmed to cruise the bloodstream, clearing plaque from artery walls before it has a chance to build up and trigger a heart attack. Perhaps some could be programmed to repair the body's cells as they grow old.

BIONIC PEOPLE
Nanotechnology not only has the potential to keep people in a good state of health; the mini-

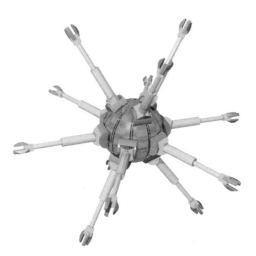

THE UTILITY FOGLET *(above) is one interpretation of what a nanobot might look like. It has a body the size of a human cell and 12 arms. A handful of such robots might form a "robot crystal" by linking their arms in a lattice structure.*

ature machinery might also be used to enhance human physiology. For example, United States researcher Robert Freitas has designed an artificial red blood cell—a tiny machine that contains even smaller pressure tanks, which carry stores of oxygen and carbon dioxide. An army of the cells could be injected into a person's bloodstream, where they would distribute oxygen throughout the body in the same way as their natural counterparts.

Each cell would be equipped with sensors and would carry an onboard computer. The cells would measure the oxygen and carbon dioxide levels throughout the body and, where the levels weren't right, would release or absorb the appropriate amount of each gas, just as natural blood cells would. But these artificial cells would be 200 times more efficient, allowing for some remarkable applications. With artificial cells in your bloodstream, you could sit at the bottom of a swimming pool for four hours without breathing, or sprint for 15 minutes without taking a breath. The applications are profound. Artificial blood cells could help victims of heart attacks by keeping their body tissues oxygenated and alive for several hours, until they were resuscitated.

ARTIFICIAL *red blood cells (nanobots) have been injected into a trauma victim's body in the artist's impression above. These synthetic cells, represented here as blue spheres, are 200 times more efficient at distributing oxygen than their natural counterparts (depicted in red). Such artificial blood cells would make it possible for the swimmer pictured below to stay under water for several hours without having to surface.*

MOORE'S LAW

In 1965, computer engineer Gordon Moore noted that every year or so, the number of tiny transistors that could be packed onto a wafer of silicon doubled. He theorized that computer power would therefore double every year or so. Now known as Moore's Law, this rule of

thumb has held ever since, with even home computers becoming out of date within about 18 months of being developed.

The question is, will Moore's Law continue to be valid? Today's microchip technology will eventually reach the point when its components can be made no smaller. This is because circuitry will become so tightly packed that the electrons that drive chips will start leaking out. It has been forecast that this physical barrier may be reached as early as 2010—long before we have nano-scale components. So, just as in the 1950s when computers moved from valve-based circuitry to microchips, if computers are to continue their march of improvement today's manufacturers will have to switch from microchip technology to something new, such as nanotechnology.

Even if the technology to drive computer circuitry into the nano-scale is not immediately apparent, nanotechnology researchers such as Ralph Merkle, at Xerox's Palo Alto Research Laboratories in the United States, believe it's likely to happen. That's because Moore's Law is driven not by technology, but by demand. Powerful market forces are pushing technology in that direction and there is no reason to believe the push will diminish in the future.

NANOMANUFACTURING IN INVISIBLE FACTORIES

Once fully mature, nanotechnology will not only produce medical miracles, it will also bring about a new way of manufacturing goods. Millions of tiny robots could work

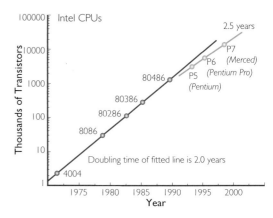

MOORE'S LAW _has stood for decades. Approximately every two years, manufacturers produce a new generation of computer processors that perform twice as well as their predecessors._

together on an invisible, submicroscopic production line. They could assemble almost anything we care to imagine, atom by atom. Once constructed, nanobots could be programmed to build more of themselves, thereby multiplying like bacteria. Throw a nanobot into a tub of raw materials and it could put together a copy of itself. Those two nanobots would then each make copies, creating four, and so on. Very quickly there would be a miniature force of millions, ready for work. But unlike real bacteria, these mechanical bugs could be programmed to stop multiplying when they have reached a certain number, or when their task is complete.

An army of nanobots could then be programmed to build the desired goods from the raw materials that are available. They might be collectively instructed to make a car engine, for instance. Not unlike a team of miniature mechanical ants working in unison, and at great speed, they would grow the engine before your eyes.

The process might seem almost magical, but there is really nothing supernatural about it: It is the way manufacturing is done in nature. A seed, after all, is essentially a pod of nanomachinery, and planting it in the ground is the equivalent of putting it in a vat of raw materials. Over months, these natural nano-

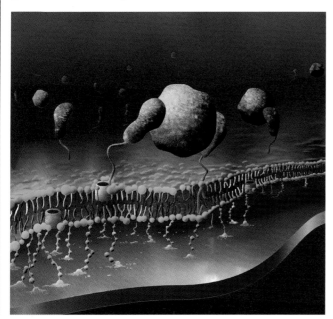

THIS BIOSENSOR _(left) acts as a biologial switch, converting a biological event into a digital signal. Applications include rapid analysis of pollutants in the environment._

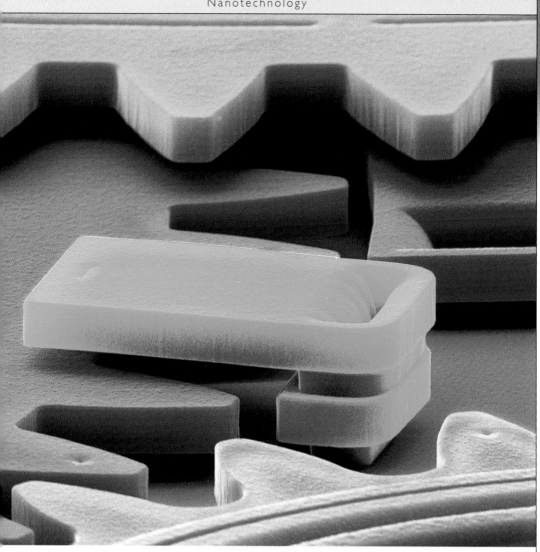

TINY GEARS *and cogs (above), where each of the cogs' teeth is about the size of a single human cell, are the basis of designing machines that are small enough to travel freely in the human body; Eric Drexler (right) is just one of the people exploring the further applications of nanomachinery.*

machines assemble a beautiful flower, atom by atom, from the dirt and other raw materials. With nanotechnology, say proponents, we'll be doing essentially the same thing. And growing our goods in this way will make them very cheap, from car engines to entire buildings.

INTELLIGENT NANOMATERIALS

Eric Drexler was one of the first to outline some of the extraordinary possibilities of nanotechnology in his seminal book *Engines of Creation*. He asks you to imagine a futuristic spacesuit. It is gray and of rubbery appearance, with a see-through helmet. You pick it up; it feels quite heavy, but once you put it on the weight just disappears. You can walk, run, jump, bend, and stretch in it with no apparent resistance at all. You rub your gloved fingertips

together and the sensations are as real as if there were no material between them. With this suit on you feel as free as if walking around naked. It's equipped with heating and cooling to keep its wearer at the perfect temperature, regardless of conditions outside. It carries an ample supply of fresh air. And the apparel even comes with food and drink supplies, and a small backpack to propel you through the vacuum of space.

OK

The material covering of the suit is really the equivalent of a solar panel, providing the wearer with the energy to run the oxygen supply, the onboard radio, and the suit's propulsion system. It could also be used to power nanobots to convert the wearer's exhaled carbon dioxide back into oxygen.

This remarkable suit is built of the kind of intelligent materials that will become possible with nanotechnology—materials that are interwoven with nanocomputers, nanosensors, and nanomotors. In essence, it is cloth that can move and think for itself. For example, the nanomachinery will be programmed to push and pull the suit so that it moves precisely as you do. That's why no resistance is felt.

Your cloth-covered fingers can feel things with as much sensitivity as real fingers because the suit has pressure sensors spread over its outside surface. When you touch an object, these sensors measure the forces, that are then transferred to your actual fingers via "pressure creators" that line the inside of the suit. Apart from your hands feeling a little thicker than normal, you wouldn't know the difference.

A suit that can be programmed to move as you move could also be instructed to amplify

THE CELL ROVER in this artist's impression (above) could swim through the human body, performing useful medical tasks such as delivering drugs, removing waste products or toxins, targeting cells, providing intracellular transport, and repairing cells. The blue, octopuslike nanobot (below) is one of billions of brain cell enhancers. The central sphere houses a computer, with a storehouse of information equal to many large libraries.

strength. For example, it could apply a force ten times the one you exert, making you powerful enough to bend steel and leap over buildings. Similarly, the garb could be programmed in the opposite way, transmitting only a tenth of the force experienced outside the suit. Falling from a building or being hit by a car would be reduced to minor sensations,

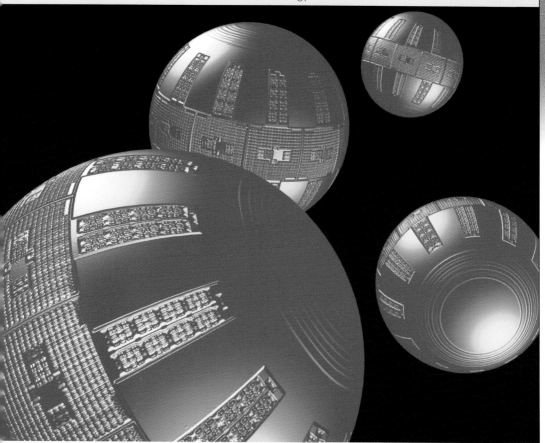

like no more than a bump. The garment could be programmed to be as strong as steel or as floppy as rubber, depending on what the user's requirements are. It could even be made to repair itself, fixing any damage as it occurs and constantly mending general wear and tear.

Everyday clothes will also be made from these intelligent materials, predicts Drexler. Put on your new shoes and they will instantly change shape, molding themselves perfectly to the contours of your feet. Clothes could vary their texture as well as their shape, keeping you comfortable no matter what the weather.

The applications of intelligent materials will be wide-ranging. Even something like painting the house could change forever. Rather than a liquid, paint will come in the form of a solid block, predicts Drexler. With a special pen, you would mark out the area on the wall that you wished to be painted and then trowel a chunk of the solid paint onto it. The paint would then "melt" onto the wall and spread itself evenly over the selected area. Inside the structure of the paint, nanomachines could later be pro-grammed to vibrate in unison, like a speaker, giving your living room walls the ability to produce music. Conversely, the paint's nano-machines could be programmed to absorb any sounds. With the walls switched into this mode a party next door would be barely detectable.

POSSIBLE ARTIFICIAL *oxygen carriers have been under investigation for eight decades. In this artist's rendering of nanobots that could fulfill the role (above), pumping station geometry and polar barcodes are clearly visible.*

THE AGE OF DIAMOND

Nanotechnology will allow buildings and cars, for example, to be made of materials that are inconceivable today. Take carbon: It's the element of choice when assembling things, atom by atom, in nature. The nanobots could be programmed to take charcoal, a cheap source of carbon, and then rearrange the atoms into a far stronger carbon product—like diamond.

It is really a matter of how carbon atoms are arranged, and placing them into a crystal lattice structure makes them diamond. And because diamond is approximately 50 times stronger than steel for the same weight, not only could a car engine be fabricated from it, but so could much of the car's body. The result would be a vehicle stronger than any seen in the 20th century, but so light it could practically be lifted into a parking space. And, quite possibly, so cheap you might not have to worry about having anyone pick it up and walk away with it, since everyone would have one, and they would be easily replaced.

Cloning

Scientists have recently produced an array of clones

from adult mammals—and the future implications for

the human species are enormous.

During the final few years of the 20th century, the concept of artificially cloning humans emerged from the realms of science fiction to become a biotechnological possibility. A rapid series of developments in the field of animal cloning led some scientists to speculate that the global human population could include an artificially produced human clone in the first few decades of the 21st century. Of course, the question of whether it should happen is another matter and one that governments, social commentators, scientists, ethicists, and religious leaders around the globe will undoubtedly still be debating long after it *has* actually happened.

A CELEBRATED SHEEP

Frenzied debate on human cloning began soon after the world first learned of Dolly, a quite ordinary-looking sheep with an extraordinary claim to fame. Dolly was the first mammal ever to be cloned from an adult. She was thrust into international stardom in 1997 when her creators, a research team from Scotland's Roslin Institute, announced that she was genetically identical to her mother.

Three different adult sheep were involved in the creation of Dolly, none of which was male. An udder cell was taken from the first, and the second provided an ovum, from which the scientists removed the genetic material. During a process known as "nuclear transfer," the udder cell and the empty ovum were fused. With a little encouragement, the fused cell began to develop into an embryo and was then implanted into the uterus of a third sheep, a surrogate mother that carried the offspring to term. After Dolly was born, tests confirmed that she was genetically identical to the first sheep—in other words, she was a clone of her mother.

Although straightforward in theory, the procedure required various forms of intervention and much practical experimentation along the way. Over 270 attempts led to 29 embryos, which in turn resulted in just one success. Even so, much of the world's scientific community was genuinely taken aback by the breakthrough. Frogs had been cloned in the 1960s, but few biologists dared to imagine it would be

IAN WILMUT *(above) led the research team from the Roslin Institute in Scotland, which created Dolly, the sheep declared in 1997 to be the first mammal cloned from an adult. Dotcom, Millie, Christa, Alexis, and Carrel (right) were cloned in 2000 at a Virginia Tech University research facility. They were the first-ever litter of pigs, following similar cloning of cattle and mice.*

DURING NUCLEAR TRANSFER, *a cell from one animal is injected into another animal's empty egg cell, held in place here by a micropipette. The resulting embryo develops under the genetic instructions contained in the nucleus of the donor cell.*

possible to generate a whole new animal using a cell from an adult mammal. Ever since Dolly, researchers have been improving and modifying the nuclear transfer technology that was used to create her, successfully applying it to a wider range of mammals. Results have included pig, mice, cow, and goat clones.

ORGAN FACTORIES

In the shorter term, cloning based on nuclear transfer will probably have its biggest impact in the area of animal production. For example, if combined with genetic engineering technologies, it could be used to produce large numbers of identical livestock that carry the same genetically desirable features. This would, of course, have huge implications for animal husbandry, but it could also lead to some quite spectacular achievements in the area of human health.

In just one of many promising applications, cloned domestic pigs could become a source of replacement body parts for diseased or damaged human organs. There is already a huge worldwide shortage of human organs available for transplants, and demand is expected to continue growing throughout the 21st century. Pigs are widely regarded as a preferred alternate source of organs. One reason for this is that their body

BACK FROM THE DEAD

Researchers at the Australian Museum are investigating whether cloning technologies, when combined with other advances in biotechnology and reproductive biology, could help bring an already extinct species, the Thylacine, back from the dead. Known also as the Tasmanian Tiger, the Thylacine was a carnivorous marsupial (pouched mammal) hunted to extinction during the 19th century and early 1900s by European settlers. The last known Thylacine died in a zoo in 1936. The Australian Museum, however, has a Thylacine pup preserved in alcohol, and scientists hope that one day it might be possible to extract DNA from this specimen and use it to re-create the lost species. Scientists acknowledge that the project is very complex, offers no guarantee of success, and, even if all proceeds as planned, will take at least 20 years and as much as $30 million to get a result.

parts are comparable in size to those of humans. However, the human immune system usually mounts a severe rejection reaction against transplanted pig tissues and organs. Scientists are trying to overcome this problem by combining nuclear transfer with other genetic engineering technologies to produce pigs with organs that are specifically tailored for human implantation. These organs are much less likely to be rejected than body parts from ordinary pigs.

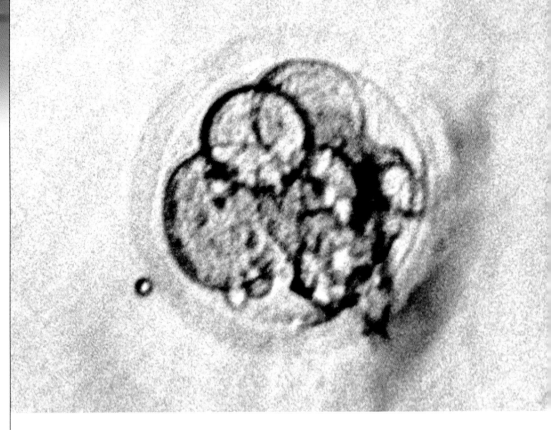

THIS MICROSCOPE PICTURE *shows the very early cell divisions of a developing embryo. This is the stage—after less than 14 days of growth—at which embryonic stem cells would be removed for therapeutic cloning.*

MODERN DAY "NOAH'S ARKS"

Nuclear transfer and techniques such as "embryo cloning"—an artificial form of twinning—could contribute to the preservation of endangered species such as the Sumatran Tiger (above). Scientists around the world have begun collecting and storing frozen tissue samples of endangered animals for which cloning could prove to be a last resort. Already, work has progressed toward cloning the Sumatran Tiger and China's Giant Panda using the same technology that created Dolly the sheep.

THERAPEUTIC CLONING

In the more distant future it could be possible to use cloning technologies to produce tissues, cells, and perhaps even whole organs that are so precisely matched to their intended recipients there would be no rejection risk at all. Consider, for example, a patient with leukemia who needs a bone marrow transplant in order to treat the condition. Researchers envisage one day being able to take a healthy cell from a patient's body (perhaps from the skin) and then reprogram it to produce new bone marrow cells with the same genetic identity as the patient.

The catch is that this would require the use of human eggs and of very-early-stage human embryos, which to many people is not morally acceptable. The process, labeled "therapeutic cloning," would work along these lines: First, the patient's healthy cell would be fused with a human ovum from which the genetic material had been removed, creating a cloned embryo equivalent to the very early stages of Dolly's development. The embryo would only be allowed to develop into a ball of cells, however. At that stage, embryonic stem cells would be removed. These have the unique ability to develop into any type of cell in the body. And if scientists could work out the controls needed to manipulate stem cells, they could, theoretically, coax them to grow into any cell that may be required. For a leukemia patient, that would

RENE MAGRITTE, *in his 1953 work Golconde (left), depicted exact copies of himself in unremitting repetition. What the Belgian surrealist achieved with the brush, scientists will never be in a position to match in the laboratory.*

adolescence—can make the final outcome from any particular set of genes extremely variable. Even identical twins have different personalities, though they were born at the same time and raised in the same environment. Often they even look slightly different. Scientists believe that, although clones will certainly bear at least a physical resemblance to their genetic donors, they will still be distinct in many ways from their donors.

The merits and faults of cloning humans are already being considered by panels of experts around the world who are charged with addressing the ethical implications of the relevant technologies. A major challenge for all societies in the years to come will be to ensure that cloning is adequately regulated in both people and animals, that in moving forward into this new era of technology, we use it for the good of humanity and not its detriment.

mean new bone marrow cells. For sufferers of Alzheimer's, or Parkinson's disease, it would be healthy nerve cells to replace damaged ones. In theory, the possibilities are endless.

DELAYED TWINS

Therapeutic cloning is at one end of the spectrum of potential applications for nuclear transfer technology in humans. At the other end of the spectrum is the production of complete, artificially produced human clones. Natural human clones—individuals with exactly the same genetic make up—already exist in the form of identical twins. But what would artificially cloned humans be like if the immense scientific challenges required to produce them are eventually overcome? Well, they wouldn't be carbon copies of the donors who provide their genetic instructions, say the experts. Rather, human clones would be what scientists refer to as the "delayed twins" of their donors.

There is a common misconception that our genes can only be "expressed" in certain preordained ways, and that our genetic destiny can follow just one path. The scientific consensus, however, is that the way in which our genes are expressed, the way they affect our outward appearance and behavior, is strongly influenced by environmental factors.

What happens to us on the journey from conception to adulthood—from the environment we experience in the womb, to the nutrition we receive as a growing child, to the social circumstances we experience through our

IDENTICAL TWINS *(right) may look very similar but often have quite different personalities and abilities. Scientists believe that this is because the way in which our genes are expressed often depends on how we react to environmental factors.*

Human–Machine Interface

*Say goodbye to your mouse and keyboard. Machines are
already learning how to read your thoughts.*

Ever since the dawn of humanity, people have used their hands to control the tools they have built. Even in the highly technological world at the beginning of this century, our hands still do almost all our interacting with objects. Whether we are pushing a button, sliding a lever, or skidding a computer mouse across a plastic mat, our finely tuned fingers and hands control the interaction with machines that enables us to drive cars, operate cash machines, and type e-mails.

But the future of human-machine interaction is slowly being taken out of our hands. The goal of much research around the world now lies in getting a machine to understand what it is that we want to do before we even express it. By the end of this century, machines might well be listening to our thoughts.

In order to make such advances and develop new interfaces with machines and tools, researchers have to step back from the way things are currently done. Interface designers are attempting to lose all their preconceptions and imagine the best way—not necessarily an improvement on the current way—for us to interact with a machine. From this process, one thing has already become clear. Buttons and levers might be around for a while, but the days of the keyboard are numbered.

They might be commonplace now, but in a few generations' time, people will look at the keyboard in the same way that we now look at the slide rule. It is an extraordinarily primitive and unintuitive way to put information and instructions into a machine. It relies on pushing buttons in exactly the right sequence, a skill that takes years of training and practice to master. It has to be re-configured depending on the user's native language. And, no matter how miniaturized advanced technology becomes, it will always be bulky and heavy because it is limited by the size of our fingers.

Technology is already beginning to move away from the keyboard, with touch screens that recognize handwriting or respond to jabs on specific icons. But an even more natural way for us to communicate with machines is the way that we often interact with other

ROBOTS ARE ALREADY TAKING *the place of humans when it comes to giving us information and entertainment. Sweet lips (above, left) is a robot guide that takes visitors through the Carnegie Museum of Natural History in Pittsburgh in the United States, talking and playing information videos. Sega's Galbo amusement park in Japan uses a robot whale (above) to enhance the experience of video gamers.*

humans: to talk. At the beginning of this century, a few examples of voice recognition software were being sold commercially. Such software avoids the need to use a keyboard to type a document but requires a significant amount of practice time for the program to recognize the user. The user also has to speak directly into a microphone: Programs are not able to filter out background noise effectively, or even the reverberations that occur when talking in the middle of a room. However, with further development, speech will soon provide our primary means of interacting with machines around us. We'll talk to the car, telling it our destination, and its technology will provide us with a recommended route.

CHIP CONTROL

Eventually, for some applications, we may not even need to speak; many of the machines of the future will be able to respond simply to our presence. Microchip implants have already enabled certain bold individuals to interact with the computers and machines in their environ-

ment. When the technology becomes more reliable and more widely available, it might authorize our use of certain computer systems or automatically open doors that remain shut to others. In the car, an implanted chip could communicate with the driving seat, adjusting it to best suit our height, weight, and posture. Implants may carry educational or reference information to be downloaded onto the nearest computer, or even our favorite music ready for downloading onto a nearby hi-fi system.

Implants are already playing an increasingly important role in medicine. Researchers are now putting microchips into the retina of people with vision problems, and helping them regain some of their sight. The chips are able to convert light into electrical signals, which are fed into the areas of the brain that interpret vision. Chips have been inserted into the bodies of those paralyzed by spinal cord injuries, chips that make muscles contract on receiving a signal. Some paralyzed patients have walked as far as a mile (1.6 km) using these implants. Eventually, implanted chips that can interpret instructions in the form of electrical signals from the brain or nervous system will enable those with paralyzing spinal injuries to walk by themselves. Make the chip respond to the patient's impulses, rather than a signal from a box, and the bionic man will be up and running.

Controlling machines through brain and nerve signals is already happening. Scientists

THE INTELLIGENT KNEE, *developed by the Massachusetts Institute of Technology in the United States, can change the world for amputees. It senses and reacts to its environment, allowing its wearer to walk up steep slopes and through snow.*

have set up a link between a sea lamprey's brain and a small mobile robot. The scientists were able to pass signals back and forth between the lamprey and the robot. Light signals from the robot triggered activity in the sea lamprey's brain, and the sea lamprey's brain was able to set the robot's wheels in motion. It was a two-way interaction between the hardware and software of the robot, and the "wetware" of biological tissue.

READING THE BRAIN

Later this century, our implants won't just be in us and interacting with the outside world. They will respond to biological signals within our body. Chip implants are already being developed that can deliver doses of medicine, such as insulin, in response to our body's chemical signals. But the ultimate goal is for implanted chips and external machines to respond to our conscious thoughts. It's not science fiction: Electrodes placed on the brains of rats have already allowed the animals to turn on a drinking water supply with their thoughts. And disabled people have controlled a computer's cursor by imagining the movement of their limbs.

But the development of fully controllable human-machine interfaces will require an intricate understanding of the operation of the human brain. Researchers are already building cortical decoders that unravel the brain's sig-

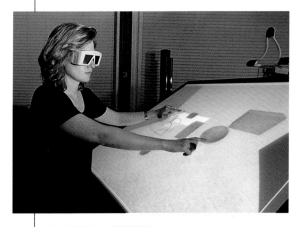

INDUSTRIAL DESIGNERS *use 3-D glasses to bring their on-screen creations to life. By touching the screen, a designer can change the component's shape, and view it from any angle, almost as if they were holding it in their hands.*

nals. These are to be implanted in the parts of the brain that control the muscles of the body, and would record the signals the brain sends out for muscle movement.

Another way to decode the way the brain works might be through the detailed mapping of the position of neurons in the brain. Neurons are the brain cells that grow links between each other as the brain develops. By cutting the brain into extremely small slices, and using computers to scan images of each slice, it will be possible to create a 3-D map of all these interconnections. From there, it should be possible to watch how the signals are sent between neurons and determine how the whole system works. But this is a huge task, and will require the kind of effort that has gone into mapping the human genome.

SILICON MIND

Once the task is done, however, it might then be possible to re-create brain function using neural networks—computers that operate in the same way as brains. Scientists are now beginning to create neural networks using ultra-small components. Some researchers believe that, if the human brain could be analyzed and understood, the function of an individual's brain could be mimicked by, or even downloaded onto, an artificial neural net.

Such advances could revolutionize entertainment and education, allowing you to make your experiences, feelings, or knowledge

KEVIN WARWICK *(above), the first human cyborg, doesn't have to touch doors. A signal from the silicon chip implanted in his arm opens the door as he approaches.*

ELECTRONIC TELEPATHY

Kevin Warwick, a cybernetics researcher at the University of Reading, England, plans to use human-machine interfaces to create a kind of telepathy. He is drawing up plans to have microchips implanted in himself and his wife. The chips will be linked into their nervous systems, and will also share signals. In theory, this set-up should allow Warwick and his wife to receive each other's brain and nerve signals.

Warwick works with neurosurgeons, attempting to improve the medical implants that can help paralyzed people to walk. He has previously had chip implants that make nearby machines respond to his presence, opening doors and operating his personal computer. It is unlikely that the electronic link between Warwick and his wife will result in them sharing unspoken thoughts straight away, but it will help give all scientists much-needed, further insight into the way the brain works.

available for others to share. But the idea raises immediate ethical concerns. How much of a person's brain has to be controlled or influenced by a machine for us to consider them less than human, for instance? And how much biological matter does a machine have to control before we consider the machine partly alive? If we give a machine the power to affect our

senses by sending signals directly into the brain, could it give us a "sensory overload" and damage the biological connections? These are just a few of the questions that will only be addressed as the technology develops.

By the end of the century we may interact with machines in extremely subtle but astoundingly complex ways. In fact, it may be hard to know whether it is us or the machines that are the true cyborgs.

MICROPROCESSOR CHIPS *implanted in the legs have helped paraplegics like Marc Merger (below) to walk again. Controlled by a computer, the chips stimulate dormant muscles to contract in the right sequence to produce a walking motion.*

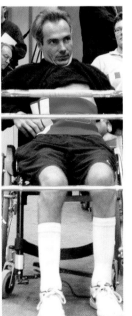

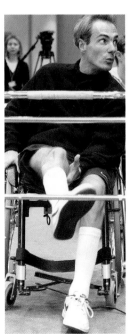

Optical Fibers

Optical fiber networks have already given us sophisticated services and information at our fingertips, but this is nothing compared with what we can expect in the future.

It is difficult to think of modern society without thinking about the information revolution. Not only is the amount of information in our society growing, but it is also increasingly being transferred all over the world, via the World Wide Web. When it comes to the transfer of information over long distances, optical fibers are hard to beat. Though one might think that information traveling through optical fibers consists mostly of telephone calls, today the majority are used for the transfer of data, for example in downloading a Web page or establishing a video link. Moreover, since the amount of data we are sharing is ballooning, while the number of phone calls that we make is increasing only slowly, soon almost all fiber-optic traffic will be made up of data.

CHANNELS OF COMMUNICATION

An optical fiber is a strand of glass that is much thinner than a human hair. Information travels through a fiber as a pattern of very short light pulses that encodes information (a telephone call, for example). These light pulses, coming from a laser, enter the fiber at one end. The fiber then guides the light along its length until it emerges on the other side, where the pattern of pulses is reassembled into the telephone conversation. For this to work properly, the fiber must be very transparent. This transparency was only achieved in the 1970s, and the glass that is now used in optical fibers is over a million times more transparent than window glass. This degree of transparency is somewhat similar to sticking your head under the water and being able to see clearly people who are swimming tens of miles (kms) away.

Optical fibers have now completely replaced copper cables in the transfer of information. The reason is simple: many more light pulses can be squeezed through an optical fiber than electrical pulses through copper cable, enabling the optical fiber to carry much more information. Optical fiber is now being laid all over the world at a total rate of more than 15,000 miles (24,150 km) a day, or 600 miles per hour (970 km/h), which is roughly the speed of sound. The capacity of an individual optical fiber has increased tremendously in recent years (though still only a few percent of the available capacity is being used). But this capacity has barely kept up with demand, or with the ever-increasing amounts of information to be sent. And, because of the novel applications described below, required capacity will need to increase for years to come.

What will all this capacity be used for and how will it affect us? An example is interactive television, which lets the viewer, whom we'll call Alice, decide when she wants to see a particular program—on African animals, say. Once she has chosen this program, she wants to skip the warthogs,

THE ABILITY TO PUT *many switches in one small place has given researchers the opportunity to explore completely new ways of manipulating light in a fiber system (left), changing how we move and manage light.*

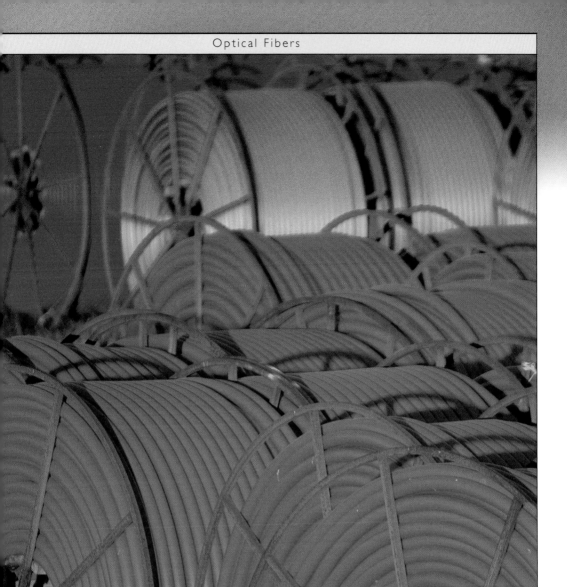

FIBER-OPTIC CABLES *(above) have now completely replaced copper cables because of their greater information-carrying capacity. Fiber-optic cables are now being rolled out around the world at a speed roughly that of sound.*

which she isn't interested in, and see more of the big cats, which she finds fascinating. But Alice's neighbors want to watch the news, and they want extra science and technology items and no sports. So the optical fiber network by which these programs are delivered must have the capacity to handle different programs for each household, rather than a limited selection from which viewers must choose.

LIGHT TOUCHES

Other applications that will require enormous capacity include interactive services such as interactive shopping, telemedicine (remote medical diagnosis and treatment), and telepresence, which would allow, for example,

for a virtual presence at a remote location and interaction with other visitors, real or virtual. With telepresence, Alice could attend a family reunion without leaving home. She could not only chat with her family and see what her uncle looks like after his recent facelift, but also, using a robotic device, hug her newborn niece and shake hands with her sister's new partner. The optical fiber network would provide not only the speech and the pictures but also the subtle nuances of the sense of touch in the handshake and the hug.

If the fiber transparency is so important, are there ways to improve it further? One possibility is to have the light travel through air rather than through glass. This is achieved in novel fiber types, in which the light travels through an air hole running along the length of a fiber surrounded by glass. The properties of such fibers will be just one of the hot topics of the science of the 21st century.

Supramolecular Chemistry

The cell is an enormously complex and powerful machine.

Taking tips from the way it works will help us build giant

molecules to cure disease and harvest energy from the Sun.

It is astonishing that anything as complex as a human being can develop from one single embryonic cell. The co-ordination and control required for the task is remarkable. This single cell must replicate itself into thousands of millions of other cells, of hundreds of different varieties. Along the way, it creates billions of complex molecules such as proteins and sugars, and makes them react at precisely the right moment. And it all happens unassisted, set in train by instructions coded into the structure of the chemical we call DNA.

The single embryonic cell can be thought of as a complex nanomachine, and we can learn a lot from it. By looking at the way these cells take atoms and molecules and organize them into complex structures like cell membranes, for example, we can learn to build giant chemical structures or "supramolecular assemblies" that mimic the way the cell works. These artificial cells are extremely useful. They can act as microreactors—tiny test tubes capable of creating novel chemicals. They can even be used as "guided missiles" that will fight disease in new ways.

THE DEVELOPMENT OF A CHICKEN *inside the egg seems to unfold automatically (below). The millions of molecules that make it happen are programmed to react in just the right way, at just the right place, and at just the right time. It is one of nature's perfect examples of a self-assembly system.*

One of the simplest ways we build complex assemblies of large molecules relies on almost the same chemistry that is used by living cells. Our bodies are built from a variety of proteins, sugars, and fats, yet we are almost 70 percent water. But molecules such as fats won't dissolve in water—if you mix olive oil and vinegar, for example, they separate into two layers.

Animal cells solve this problem using special molecules called emulsifiers. Emulsifier molecules possess two distinct parts—a hydrophilic head that attracts water molecules, and a hydrophobic tail that attracts oils. Because of this unique structure, emulsifiers enable fats, oils, and water to form stable mixtures.

TINY CHEMICAL PLANTS

Add emulsifiers to a mixture of oil and water and they will line up at the interface between the two liquids to form a sheet—the hydrophobic tails facing the oil layer and the hydrophilic heads facing the water. If you curve this sheet round to form a ball, you have a structure called a micelle that resembles a simple cell with a membrane. The body has many uses for micelles: To stabilize fatty molecules that are dissolved in the bloodstream, for example.

But scientists have learnt to use the same trick to form artificial "cells" that can carry out all kinds of useful chemistry. If you create a micelle with hydrophilic groups on the out-

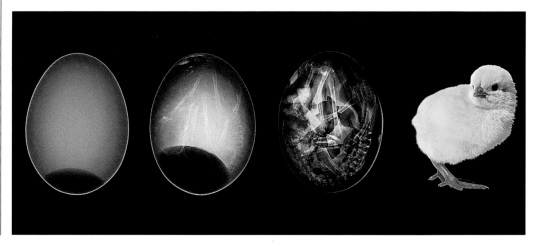

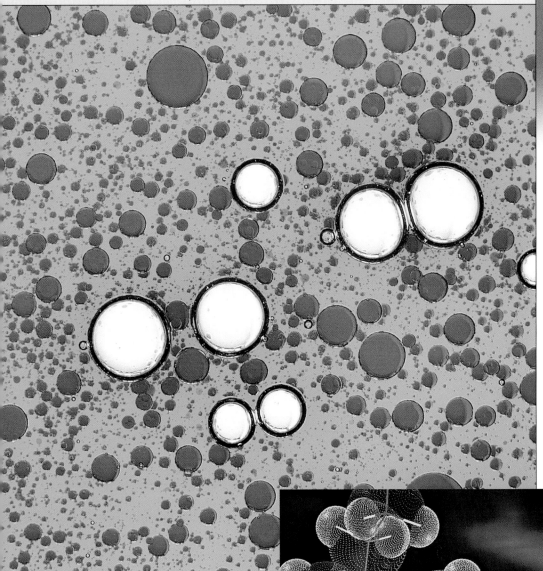

side, it can dissolve in water, yet the hydrophobic groups on the inside can trap oil-soluble molecules within the micelle. This means these artificial cells are perfect for processing substances that don't usually mix.

For example, chemists would like to use enzymes to make drugs and other pharmaceutical products. Enzymes can be very powerful biological catalysts, but they usually function in water. However, many of the compounds that chemists would like to turn into products, such as steroids and fats, are insoluble in water.

The answer is to dissolve the compounds in an oil-based solvent and then to add micelles with water-filled centers packed with enzymes. The enzymes can move to the walls of the micelles and react with the components dissolved in the oil outside.

Using micelles as chemical labs can also give scientists a way to control reactions. For example, tiny particles of semi-conducting materials

OIL AND WATER DON'T MIX *(top) because water molecules (blue) are electrostatically charged while oily molecules (orange) are not—and they each stick to their own kind. Animal cells use molecules called emulsifiers to overcome this problem. Emulsifiers (shown above, as part of a cholesterol molecule) have a unique structure that enables them to mix with both water and oil molecules to form stable mixtures.*

123

such as cadmium sulfide are important as catalysts in the chemical industry. To be useful, however, the size of the cadmium sulfide particles needs to be controlled very precisely. Micelle micro-reactors provide chemists with a simple way to do this.

The chemists create two sets of micelles, one holding cadmium ions and the other set holding sulfide ions. When the two sets are mixed, the ions react together and particles of solid cadmium sulfide form. Micelles can be made in a variety of sizes, sometimes as small as a nanometer (one billionth of a meter) across. And because the size of the particles depends on the size of the micelles, chemists can tailor the results they want. The same trick can also be used to control the size of synthetic polymer molecules such as polyacrylamide.

GROWING CHEMICAL TREES

Chemists are learning to build many other kinds of supramolecular assemblies, too. For instance, in the early 1980s, Donald Tomalia and his colleagues at the Dow Chemicals Company in the United States began to grow highly branched, tree-like polymer molecules that have come to be known as "dendrimers" (from the ancient Greek for tree—"dendron.")

Unlike normal polymers that build up in a linear fashion to form long, thin, spaghetti-like

molecules, dendrimers are made by attaching branched molecules such as amines to a central molecular core. Adding still more branched molecules, layer upon layer, creates a huge ball-like structure.

Dendrimers can be built from many different kinds of molecules: carbon-based molecules such as sugars, silicon-based molecules, amino acids, or a combination of organic and inorganic components. By precisely controlling the number of shells that wrap the core, chemists can create dendrimers of specific sizes—they can reach many hundreds of nanometers across. Since chemists can also control the chemical make-up of each layer, they can design dendrimers with specific tasks in mind.

For instance, dendrimers can be created with large clefts or pores within them. Just as with micelles, these hollow dendrimers can trap and hold atoms or small molecules such as metals or small sugars. By holding them close together, these molecules can be made to react with each other. Then, by altering the environment outside the dendrimers with a few drops of acid, for example, the dendrimers can be triggered to expand slightly, releasing their molecular cargo. This provides the potential to grab, react, and release molecules on demand, thereby creating the microscopic equivalent of a chemical factory. Dendrimers could eventually offer us new ways to detect and break down chemical pollutants in the environment or to make highly selective chemical filters to extract and concentrate valuable chemicals during industrial processing.

TINY PARTICLES OF CADMIUM SULFIDE *(below) are important because of their use as catalysts in the chemical industry. But they are only useful if the size of the particles is precisely controlled, which is where micelles come into play.*

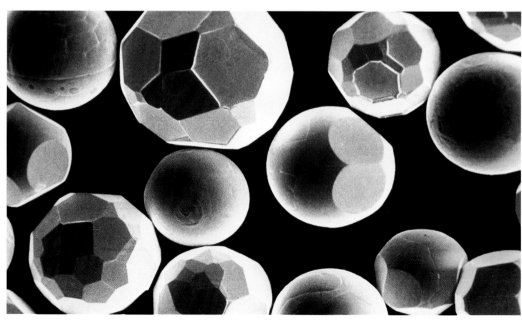

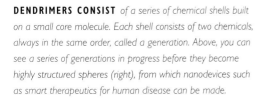

DENDRIMERS CONSIST *of a series of chemical shells built on a small core molecule. Each shell consists of two chemicals, always in the same order, called a generation. Above, you can see a series of generations in progress before they become highly structured spheres (right), from which nanodevices such as smart therapeutics for human disease can be made.*

Dendrimers could even be used to collect solar energy and convert it into electricity. To do this, you need dendrimers built from light-sensitive molecules. When they absorb sunlight, the molecules pass the energy from sunlight to the dendrimer's core where it creates electrical energy by releasing an electron.

CHEMICAL MISSILES

We are even discovering that dendrimers may offer us new and powerful ways to treat diseases such as cancer. James Baker, the director of the Center for Biologic Nanotechnology at the University of Michigan is leading a program to design giant dendrimers that will act like guided missiles when they are injected into the human body, hunting down bacteria or cancerous cells and delivering a deadly chemical payload designed to destroy the target.

This task is a huge challenge. To succeed, dendrimers must not only be capable of distinguishing and destroying any diseased cells they find, but they must also evade the body's own defenses, and then report the outcome of the treatment back to doctors.

Baker's solution is the "tecto-dendrimer," a grouping of four or five dendrimers that are bound together chemically into a single unit. Each of the component dendrimers is responsible for a specific task. One is the disease sensor, capable of recognizing unhealthy cells. This dendrimer may be coated with antibodies that identify particular kinds of bacteria, viruses, or the proteins released by cancerous cells. Once this sensor recognizes the target, it binds to it. Then the guided missile delivers its deadly load.

This comes in the form of the drug delivery dendrimer, loaded with antibiotics or chemotherapy agents. Finally, so that the doctors can follow the progress of their treatment, one component of the tecto-dendrimer contains fluorescent molecules that can be picked up under the microscope, or special metals such as gadolinium that show up during magnetic resonance imaging scans. The great advantage of the tecto-dendrimer system is that by combining different "bomb-loads," doctors can tailor the drug-system to treat many diseases.

Clearly we're still a long way from mastering the cell's intricate chemistry. However, we've already learnt enough to appreciate the tremendous potential yet to come.

MEMS

The age of micromachines has only just begun. MEMS are here. Stand by for talking labels on medicine bottles and micropumps in our bloodstream.

What started with the creation of the first wristwatch by the now-famous French clockmaker Louis-François Cartier in 1904, and continued with the invention of the transistor and the advent of the portable radio, has now become all-pervasive with the miniaturization of the computer chip.

The irresistible force towards smaller, cheaper, more powerful processors and the race to squeeze as many electronic devices as possible on to a standard silicon chip has led to huge advances in the way chips are engineered. This, in turn, has given rise to chips that are small, cheap, and reliable enough to be incorporated into almost anything—not just newly invented electronic devices like CD players, videos, and microwaves, but also more traditional items such as car engines and talking dolls.

With mastery of the manipulation of silicon for chip production, it was not long before researchers were considering whether the same fabrication techniques could be used to produce very small machines as well as computer processors. For example, Wen Ko, an engineer at Case Western Reserve University in Ohio, in the United States, worked with a team at Goodyear to design and patent tiny pressure and temperature sensors made from silicon. These sensors can be incorporated into the wall of a tire during manufacture, along with a radio transponder and an identification chip. A radio signal from a handheld scanner provides enough power for the transponder to run the sensors and send back the tire's vital statistics. In fact, this smart tire tells drivers when it needs replacing or inflating.

Such machines usually combine some way of measuring their surrounds—a sensor—with a processor to analyze the information gathered by the sensor. The results are passed on to a further device, known as an actuator, for action. The whole system is manufactured in the same way as a traditional silicon chip, which means that, once designed, these machines—or microelectromechanical systems (MEMS)—are cheap to mass produce.

MEMS are useful precisely because they are small, and are little affected by the usual forces of gravity and inertia. They can be used in arrays, and placed in small spaces inside appliances, or even living organisms. And they are becoming very common.

Perhaps the most widespread MEMS system is the one inside the head of an ink-jet printer, squeezing out the very fine droplets of ink that produce the print on a page in response to electronic impulses from a computer. The car industry now also uses MEMS to measure things like pressure, temperature, and acceleration and act where necessary. It is a MEMS system incorporating an acceleration sensor, for instance, which triggers the safety air bags in a car in the event of a collision.

MACHINING MICROMACHINES

The components of MEMS systems are produced, usually on silicon chips, by a whole series of clever surface micromachining techniques. Some techniques take their cue from microelectronics, where electric circuits are

THIS TINY MP3-PLAYER *can provide hours of entertainment anytime, anywhere. Soon, MEMS systems will be built into what we wear—glasses with video screens, for instance, and shirts with speakers, powered by fabric that generates electricity from light, heat, and movement.*

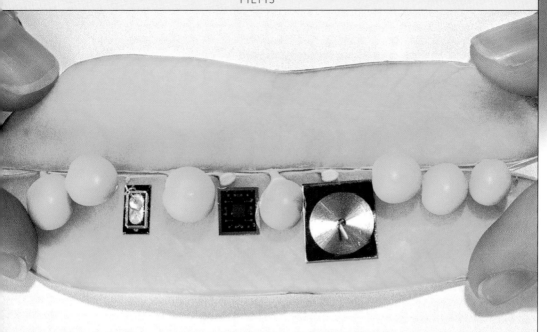

produced by etching or laying down thin layers of materials. Others make use of scanning lasers that are pulsed to bore through or cut away material into precise holes and grooves.

Using these techniques, a whole array of devices can be produced—little fans the size of a chip to cool your computer, tiny electric motors to power micropumps that may one day be implanted in the bloodstream, and little silicon beads, the size of a pin head, that could deliver insulin to diabetics.

One example of the kind of product now possible is the lab-on-a-chip developed at Scandia National Laboratories in California. About the size of an electronic organizer, the lab-on-a-chip can separate and analyze the gas or liquid components of complex chemical mixtures in less than a minute. In other words, it can perform the functions of a standard room-sized chemical laboratory quickly and cheaply on the palm of your hand. And that means it can be carried in a pocket and used to analyze a stream for signs of pollution, a crime site for traces of gunpowder or blood, or the food in a restaurant for contamination.

THE THREE PRINCIPAL *components of the micro chemical lab from Scandia National Laboratory in California can fit easily inside a snow-pea pod. Together they can collect, concentrate, and analyze chemical samples weighing less than a bacterium.*

Samples, which can weigh less than a single bacterium, are drawn into the device by a micropump and concentrated on a sponge-like structure. As the sample mixture moves from the sponge down a series of channels, the constituent chemicals travel at different speeds as they react with the materials in the channel walls. As a result, different compounds emerge from the channels at different times and are analyzed by microsensors, which include ultrasound devices, electrochemical detectors, and laser-induced fluorescence devices. The results are calculated using a microcomputer and displayed on a screen.

QUICK, SENSITIVE, *and selective chemical analysis is what the lab-on-a-chip instrument offers, seen here as cutaway artwork. Freed from the lab, researchers envision such automated field-portable systems of the future will provide results in near real time. Amongst other things, they will be able to sniff out landmines and biological hazards.*

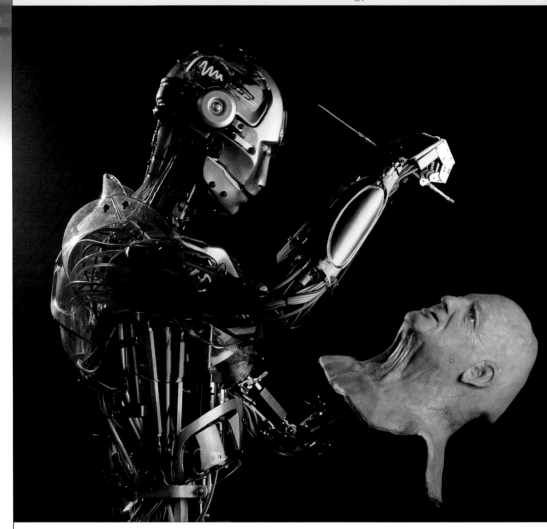

Living Materials

Scientists are giving inanimate objects capabilities to feel,

sense, and respond to the world around them within the new

and exciting field of living materials.

As the name suggests, living materials draw inspiration from the biological world. They can be used to build systems or structures capable—like living organisms—of perceiving and reacting to changes in the surrounding environment.

The field is very new, but already scientists are talking of boat hulls that change form in response to vibrations; airplane wings that adapt their shape without using flaps; and car seats that adjust themselves to their occupants' optimal comfort requirements. Also envisioned are toilets that can detect health problems by analyzing urine; artificial muscles, stronger than our own, that could be used in medical implants;

and concrete structures, such as bridges and dams, that can detect and heal their own cracks.

In the more distant future, it may be the field of living materials that ultimately gives us the humanoid-type robots imagined by many science fiction writers. It is at that sophisticated level of application that the term "living" will start to be most relevant. For now, however, these materials and their applications are more widely referred to as "smart" or "adaptive." When put together in a particularly sophisticated way, the description "intelligent" might sometimes apply.

The area's burgeoning promise arises from a strong multidiscipline approach, drawing on

MATERIALS THAT SEEMINGLY STRADDLE *the line between biology and engineering might one day be merged with systems that mimic the extraordinary consciousness of Homo sapiens to produce android-type robots. Will these robots have smart skin and bones, with self-repair abilities?*

the technologies, knowledge, and advancements in a range of sciences—from chemistry, physics, computer science, and engineering to biology, biotechnology, and nanotechnology.

Much of the early work on smart materials was carried out within the military. Excitement about civilian applications began to grow in the mid- to late 1980s, when aerospace firms, universities, and other scientific institutions began developing dedicated research programs. Now venture start-up companies based on commercially promising technologies are emerging.

At their most fundamental, smart structures incorporate either sensors or actuators (devices which create movement). At a higher level, systems contain both sensors *and* actuators. Even more sophisticated structures incorporate a sort of control mechanism that directs the material to respond to outside stimuli in particular ways. Smart materials truly become "intelligent" when, combined with computer technology, they have the ability to interpret and make "their own" decisions and respond appropriately to changing environmental cues.

Among the best known and most widely used smart materials are piezoelectrics. Like muscles twitching in response to nervous stimuli, piezoelectrics convert electrical energy into mechanical force. They also work in the other direction, converting movements (such as vibrations) into small but measurable electrical currents. Because of this dual ability they are often employed as both sensors and actuators within the same structure.

Birthday cards that "sing" when opened are among the simplest commercial products available with piezoelectrics. Smart snow skis with built-in piezoelectrics that sense, oppose, and cancel out vibrations are a more advanced application. In the automotive industry, piezoelectrics are already being used in high-precision car shock

RESEARCHERS are developing "smart dust," tiny computerized sensors that could perform a myriad of tasks from collecting meteorological data or monitoring the internal workings of an insect to finding clues at a crime scene.

absorbers. Smart paneling embedded with piezoelectrics could, in the not-too-distant future, give cars and planes enormous capabilities to sense and respond to prevailing conditions. When used as actuators, piezoelectrics show promise in the development of ultra-high-fidelity stereo speakers and could, in fact, be one day incorporated throughout entire car interiors or house walls to effectively turn those structures into speakers.

Another popular group of smart materials is shape-memory alloys (SMAs). These metallic materials always "remember" their initial shape. An SMA can be distorted and then, at a certain temperature, made to return to its original form, producing motion in the process. SMAs are actuators that work more slowly than piezoelectric actuators but are capable of larger movements. One way of exploiting an SMA's movements is to incorporate it into a structure containing a material that will resist the motion. In this way, an SMA can exert an ongoing but also controlled force. When this technology is applied to orthodontics, for example, dental arches containing a nickel-titanium SMA will provide a more steady and persistent force against the teeth than traditional steel arches. Nickel-titanium SMAs are also undergoing trials in robotics because they can simulate the steady movements of human muscle.

Electrorheological and magnetorheological fluids are smart materials that contain microscopic particles suspended in oil. When strength is needed, electric or magnetic fields can be applied to align the particles into chains or filaments, almost solidifying the liquid. And, alternatively, when flexibility is required, they can be returned to a more liquid state.

Superconductors

Cheaper and cleaner electricity, smaller and faster computers, and levitating trains are just a few of the ways in which superconductors could transform the 21st century.

In March 1987, thousands of scientists tried to squeeze into the ballroom of the Hilton Hotel in New York for a landmark meeting hosted by the American Physical Society. Many of them spilled into a nearby hallway, forced to watch proceedings on television screens as the nightlong event took on such proportions that it would later be dubbed the "Woodstock of Physics."

Up for discussion were the latest findings on superconductors, materials that can carry electricity without any resistance. A flurry of activity in laboratories around the globe during the preceding months had produced significant leaps forward in the field, advancing the promise of a 21st century revolutionized by these "perfect" conductors.

It wasn't the first time scientists had been inspired by the potential of superconductivity. The phenomenon was initially observed in 1911 by Dutch physicist Heike Kamerlingh Onnes, who found that when mercury was cooled to around 4° Kelvin (-452°F/-269°C) the metal's electrical resistance suddenly and completely disappeared. Below this point (now often called the critical temperature), mercury, unlike any conductor known before, could

carry an electrical current without any loss of energy. The possible implications for the electricity industry alone were enormous.

In the following seven decades, many other superconductors were discovered, but all, like mercury, needed to be chilled to critical temperatures so low they could only be achieved using the expensive and awkward-to-use coolant, liquid helium. This severely restricted the application of superconductors beyond the laboratory. One notable exception was magnetic resonance imaging (MRI). This medical technology, which allows the internal workings of a body to be seen without surgery, has used superconductors since the late 1970s.

THE WAY AHEAD

The breakthrough that precipitated the celebrated New York gathering came in 1986, when IBM researchers Georg Bednorz and Alex Müller reported on a new ceramic material they had produced that exhibited superconductivity at a temperature significantly higher than anything before. Laboratories in Japan, the United States, Europe, and China scrambled to take the work further and rapidly developed superconductors with critical temperatures above 77° Kelvin (-321°F/-196°C). These "high-temperature" superconductors could be produced using liquid nitrogen, a cheap, plentiful, and easy-to-manage coolant. It opened the door for the development of many more practical applications.

Superconductors with even higher critical temperatures have since been produced, and the search continues for a room-temperature superconductor that could put this technology into routine use in our homes. Even without it, however, superconductivity is already destined to leave its mark on this century.

Medical technology, materials processing, transportation, electronics, and computing are just a few areas in which commercial uses are being developed. The most wide-reaching impacts, however, are likely to be in the power industry. Because superconductors have zero

SUPERCONDUCTORS REPEL *external magnetic fields, a property that is being exploited in the development of high-speed magnetic levitating trains. Magnets brought close to a supercooled superconductor (above) appear to float in mid-air.*

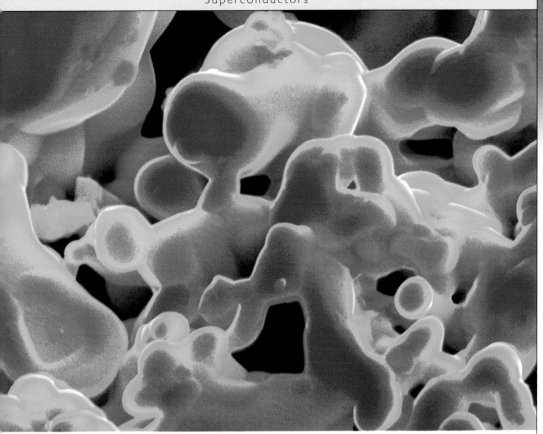

resistance, they transmit electricity without any significant loss of power, unlike traditional conductors such as copper. With nonrenewable energy sources (such as coal) fast running out, we need to use what is left more efficiently, to make it last. For this reason, many scientists believe the power industry will, during the early part of this century, have to overhaul and upgrade electricity generation, transmission, and storage with superconductors. Superconducting generators would be far smaller and far more efficient than those presently available, and superconducting transmission lines would not suffer the high rate of energy leakage that is experienced at present.

With superconducting electricity storage systems, power companies would have much improved capabilities to manage electricity supply and would cope better with fluctuations in demand. Such storage systems would make clean, alternative power sources such as wind and solar energy far more attractive to the power industry. Use of such sources is presently limited because their

DR. PAUL CHU (right) believes a room-temperature superconductor will be found; most theories do not exclude the possibility.

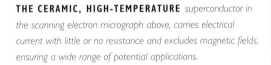

THE CERAMIC, HIGH-TEMPERATURE superconductor in the scanning electron micrograph above, carries electrical current with little or no resistance and excludes magnetic fields, ensuring a wide range of potential applications.

supply of electricity fluctuates and does not always coincide with peak demand.

Superconductors could also have a significant impact in communications and electronics. Researchers have already made a transistorlike superconducting device that does not produce heat or leak electricity like standard transistors do. In addition, superconducting electronics are being explored in the search for a "super switch" to handle the Internet's escalating rate of data traffic. The capabilities of conventional technology to handle the growth of the Internet are limited. Superconducting devices could lead to faster computers with enhanced capabilities for data storage.

As well as exhibiting zero electrical resistance, the other key attribute of superconductors is that they form perfect diamagnets so that they repel magnetic fields. Magnetic levitating trains that appear to float above superconducting tracks are among the most spectacular applications of this.

> *It is impossible to travel faster than the speed of light, and certainly not desirable, as one's hat keeps blowing off.*
>
> WOODY ALLEN (b. 1935), American film actor and director

CHAPTER FOUR

TRANSPORTATION

The Automobile

We could soon be witness to the most extensive metamorphosis

of the car since its beginnings.

Nikolaus August Otto started the idea rolling in 1878 when he engineered the gasoline engine to transport a wealthy few. Henry Ford made it available to the masses thanks to mass production principles, and the creation of the 1908 Model T, which was simple to build and operate.

The two concepts were truly milestones in history, but so well has the automobile been honed that it's easy to take for granted its outstandingly high levels of performance, comfort, convenience, durability, and occupant safety, achieved at an "affordable" cost. Even its major drawback—exhaust emissions and the resultant urban smog—is close to being eliminated.

But as the popular automobile approaches its 100th birthday, its very existence and the remarkable level of personal mobility that it affords us are at risk. This personal mobility is predicated on the ready availability of gasoline, but this century will see oil, gasoline's source, becoming scarce and much more expensive. That we often pay less now for gasoline than we do for bottled water will soon be seen as unbelievable. And while the most toxic of the car's exhaust emissions have been removed, the eventual elimination of the automobile's emission of the greenhouse gas, carbon dioxide, is a political imperative.

As the automobile becomes affordable for additional hundreds of millions of people in developing countries, such issues will increase in significance. Thus, the challenge of producing automobiles which solve the fundamental energy and greenhouse problems that are associated with automobiles while retaining near-current levels of mobility and safety is very much open.

THE MODEL T *(left) established the popularity of the petrol engine in the early 1900s. A century on, alternatives to petrol engines are being sought.*

IN THEIR PRESENT *form, solar-powered cars like the Honda "Dream" (left) which competed in the Darwin/Adelaide World Solar Challenge in Australia, are too impractical for conventional transport. Another alternative fuel is methanol (far right of right) which burns far cleaner than gasoline (far left of right).*

THE HYBRID VARIETY

In the short term, power units with greatly improved fuel economy will help to conserve oil and reduce carbon dioxide emissions. A fuel economy figure of 80 miles per gallon (3 liters per 100 km) is seen as necessary and is already achievable. But the small engines that provide this economy do not produce the power that's required for brisk acceleration.

Electric motors can provide efficient acceleration, but for acceptable range require heavy batteries with long recharge times.

The hybrid power unit—a gasoline engine linked to an electric motor—is a possible solution, delivering economy and performance. Most major manufacturers have demonstrated experimental models and Honda's Insight and Toyota's Prius are already on sale. All models use computer-controlled systems to optimize the use of the engine, drawing on a relatively small battery for short periods of acceleration and low-energy operation.

ALTERNATIVE FUELS

In the long term, alternative fuels promise to eliminate the automobile's reliance on oil and reduce and eventually eliminate greenhouse gas emissions. The goal is a range of renewable fuels which produce little or no carbon dioxide and which can be readily distributed—preferably by utilizing much of the existing gasoline distribution system.

Fuels "cleaner" than gasoline do exist. Liquefied petroleum gas and compressed natural gas, while not renewable, will extend the total availability of oil-based fuels. Both ethanol (made from crops such as corn) and methanol (made from biological wastes) are renewable, but still produce carbon dioxide and require large, cleared land areas. Such fuels will, however, have specific applications in rural areas.

Solar power has now proven its significance. The Australian Solar Challenge competitions demonstrated that specialized cars can travel 1,875 miles (3,000 km) in four days on solar power alone. But in their present form, solar cars are impractical for conventional use, and the technology could only be used for secondary applications like battery charging, cabin cooling and, increasingly, telecommunications.

Hydrogen is widely regarded as the fuel of the future, producing only water when burned, either in conventional engines or in fuel cells. The required energy to produce it can be obtained from renewable resources such as hydro, solar, and wind energy, and biological or photobiological technologies which use plants that convert sunlight to energy.

The widespread use of hydrogen will require improved storage and handling technologies and the adoption of efficient usage practices. On-vehicle tanks have yet to be perfected. Hydrogen, stored as a liquid, must be cooled to minus 423°F (minus 253°C), requiring energy to cool it in the first place and special materials to then contain it. Demonstration systems bleed off a small proportion of the stored hydrogen to maintain the low temperature.

DRIVING THE HYBRID POWER UNIT

Hybrid vehicles offer a novel driving experience. The engine, with an automatic start-stop feature, need not operate during start-up and low-speed driving, and engine speed varies with load rather than vehicle speed. No gearshift points are felt or heard, and during braking, retrieved energy recharges the batteries while the engine idles or even stops. Overall operating convenience is little affected, although the requirement for fuel economy will probably eventually eliminate higher-performance cars.

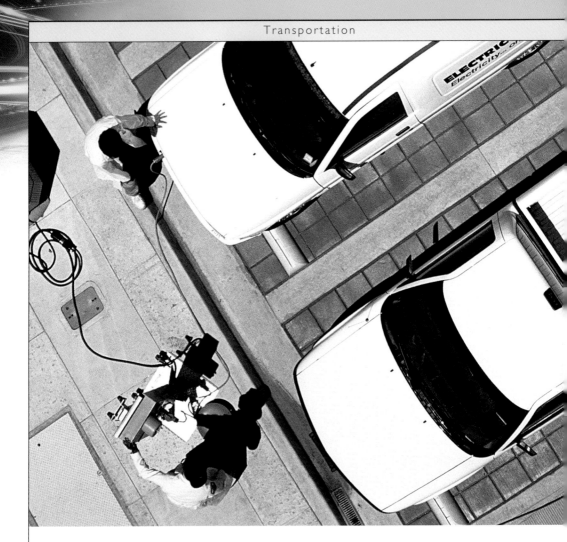

ELECTRIC CARS *(being refueled, above) are the nearest thing to nonpolluting transportation. But important issues remain to be addressed, such as how the electricity itself is generated and how adequate the batteries are.*

Storing hydrogen as a gas uses less energy, but also requires a storage medium. Car companies are currently researching compounds that will carry hydrogen at high-energy density, release it as a fuel, react quickly, and be cost-effective. Models are expected to be in fleet operation by 2005.

OPPORTUNITIES

While personal mobility will involve more considered behavior on our part, the automobile itself will be a highly refined and desirable thing to own. The digital age offers unlimited possibilities for improvements in comfort, convenience, communication, and entertainment applications.

Active suspension, seat, and noise suppression systems will further improve ride and vibration qualities. Voice-control for navigation, communication, and car control systems will make these truly hands-free for the driver, while directional audio systems will allow each occupant to interact with separate information or entertainment channels. Database-linked, predictive monitoring systems will anticipate malfunctions and advise preventive actions. These advances in convenience have their limits, however. Over 25 percent of all car accidents have been attributed to driver distraction, often involving information-age technologies. Current studies will probably define a limit for such in-vehicle equipment.

INTELLIGENT VEHICLES

The digital age is also advancing automobile safety, in particular in the area of intelligent vehicles. Until recently, the primary controlling technology for our transport systems was the four-way traffic signal. But affordable computer chips and sensors in the 1990s made possible intelligent vehicle (IV) technologies which aim to address the problems of driver error that cause 75% of all accidents. IV technologies aim to warn drivers of dangerous situations, recommend preventive actions, and even assume partial control of a vehicle to avoid a collision. They can detect the onset of driver drowsiness, warn of changes in road surface conditions, anticipate skidding, roll-over, and front-to-rear collision situations, and detect inadvertent lane deviations. They will form part of an integrated inter-modal transportation system, carrying people and goods far more safely and efficiently.

This century will see the widespread application and further development of IV technologies, some of which were demonstrated on concept cars in the final years of the 20th century. The Mercedes VITA II project, for instance, used multiple sensors to demonstrate that a vehicle can safely drive autonomously for extended periods in general traffic.

CONVERGENCE OF TECHNOLOGIES

The personal mobility provided by the automobile has come at an enormous cost—urban sprawl, traffic congestion, pollution, and accidents. These problems call for the development of integrated transport systems in which public transport vehicles have vastly improved flexibility and the automobile becomes an integral part of our personal transport system rather than our stand-alone transport vehicle.

To meet this challenge, the Bishop Austrans system in Australia has applied automobile technology to rail transport, with test tracks now in operation in Sydney. Driverless nine-passenger, van-sized cars operate six seconds apart at 75 miles per hour (120 km/h). They use a network of guideways with high-speed flexible switches, allowing cars to exit and to enter the network without affecting adjacent cars. Special bogies provide fast acceleration and braking worthy of an automobile, while a unique control system allows passengers to

FUEL CELLS

In the 1830s, fuel was produced by chemically combining hydrogen and oxygen. The 1960s space program developed fuel cells with membrane technology to do this more efficiently, allowing direct connection to electric motors without the large batteries needed in existing electric vehicles. The first fuel cells in automobiles may not, however, use hydrogen stored on board. Liquid fuels such as methanol and gasoline can be converted to hydrogen in an on-board "reformer," and this is seen as being the interim phase towards using fuel cell vehicles carrying only hydrogen.

The electric motor technology which will emerge for the hybrid power units will lead to that required for use with the fuel cell. Fuel cells will probably replace a large proportion of the conventional engines in the first quarter of this century. By the middle of this century, most automobiles will be running on hydrogen, using fuel cells. All energy will be expensive and the cost of fuel will be a major consideration in automobile purchase and usage. High-speed, long-distance travel by two or three people in heavy, powerful, automobiles will be very expensive and will be far less common than it is today.

enter any station, "call" for a car to any destination, and be collected by the next available car.

Cars stop only when they have a passenger to collect or to drop off. Energy use is vastly reduced by minimizing the weight of individual vehicles and operating only the cars that are needed at the time. Passenger capacity at peak times is equivalent to current train systems, but with much more flexibility.

Systems such as these, and others being developed through various industry and government programs, offer the prospects of acceptably private and adaptable transport. At the same time, such systems go some way towards addressing the pressing needs of energy conservation and reducing the pollution pressures on the atmosphere.

PHYSIOLOGICALLY CORRECT *airbag crash dummies, "intelligent" airbag technologies that deploy airbag movement and, of course, external airbags, will make future driving safer.*

Rail Transportation

For more than 200 years rail has been the cheapest way to move large amounts of freight over land and transport large numbers of people. But what of the future?

Most railroad lines were designed more than 100 years ago for steam-hauled trains. Nineteenth-century engineers built solidly, but frequently to dimensions that restrict us today. We have the technology now to build more powerful locomotives to haul huge loads at unprecedented speeds, but many historic lines, bridges and tunnels have to be rebuilt and tracks expensively realigned to take full advantage of this.

There are two possibilities for the future of rail transport. We can continue to develop the established railroads for longer or faster trains, and we can build entirely new lines for new technologies, some of which are very different from "railroads" as we know them today.

THE HISTORY OF SPEED ON RAILS

Before rail, freight was moved on canals or by animal-hauled carts. It took more than a week for a ton (tonne) of coal to move 200 miles (320 km) on a canal but just six hours, or less, by rail. A train first exceeded 100 miles per hour (160 km/h) in the early 1900s. In 1938 the British Mallard (above) reached 126 miles per hour (203 km/h), a still unbroken record for a steam-hauled train. Under test conditions a *TGV* has reached a record speed of 320.3 miles per hour (515 km/h); on scheduled services, they regularly exceed 200 miles per hour (320 km/h).

NEW SOLUTIONS ON OLD TRACKS

Containerization and intermodal traffic have revolutionized the handling of railroad freight by integrating rail, road, and sea transport. Freight trains with containers stacked two-high are now a familiar sight in North America and Australia; others carry complete, loaded road trucks (in Europe such trains are known as the "rolling motorway").

Very long trains are not easy to control. Coal trains in the United States and South Africa and iron ore trains in Australia can be nearly two miles (3 km) long. They can be negotiating up to four curves in different directions all at the same time, while the front and rear of the train are climbing hills and the middle is descending a gradient. Train control is often achieved by spreading locomotives along the length of the train, connecting them by radio links and operating them by computers.

AUTOMATED SYSTEMS

By 2000 there were around 80 automated rail transport systems operating worldwide. The most significant of these systems shows that driver-less operation of high frequency, passenger-carrying trains is not only feasible but also offers great advantages in flexibility of operation. Such systems include the Docklands Light Railway in London, which services 33 stations; Line 14 (Météor) of the Paris Métro; and lines in Kuala Lumpur and Tokyo. Many more urban railroads—newly built or conversions of existing lines—are likely to be fully automated in the future.

VERY FAST TRAINS

Very fast express passenger trains such as the *Train à Grande Vitesse (TGV)* of France, the *Inter-City*

Express *(ICE)* of Germany, and the *Shinkansen (Bullet Train)* of Japan are high-performance electric trains. These very fast trains run on standard rails and have stream-lining to reduce drag against the air.

Many new sections of line have been built to allow these trains to reach speeds of over 200 miles per hour (330 km/h). The new lines are much straighter and are graded to suit such very high speeds. To allow faster running on traditional railroads, trains have been constructed with tilt mechanisms that roll the train into corners, creating the effect of a steeply pitched curve, as seen on racing tracks.

However, all these very fast passenger trains have a limitation: wheels on rails. Wheels and axles create friction that not only slows the

THE JAPANESE BULLET TRAINS *lined up at a station in Tokyo, Japan (above) and the Docklands Light Railway in London, England (below), share the same dependency on the friction between steel wheel and steel rail. But ultimately this friction limits the speeds that can be attained by wheels on rails, so for the very fast trains of the future we will almost certainly have to look ahead to some new types of propulsion.*

movement of the train but also generates heat and vibration. Reducing the number of wheels on a train can minimize these problems and can be achieved by "sharing" the wheels between carriages where a common bogie supports two carriages. Land transport that is faster than the current crop of very fast trains may abandon wheels altogether and propel the train along some new kind of track.

HOVERTRAINS

There are at least two forms of transport for passengers that go faster than the current rail-bound very fast trains. The hovertrain and maglev (magnetic levitation) both suspend the train in the air above a track base. Combined with super-sleek streamlining, this almost eliminates drag and friction.

The hovertrain or aerotrain uses the ground effect principle, in which a cushion of air is created under the vehicle. The shape of the vehicle and the track achieve this, funneling air under the train to push it up as it goes along. Extension of the seagoing hovercraft concept to land through linear induction electric motors began in Great Britain, where the hovercraft was invented, but support was discontinued in the 1960s.

The French test vehicle *Aérotrain 1*, built in 1965, was propelled by ducted fan engines and reached 214 miles per hour (344 km/h). The rocket-assisted *Aérotrain 2* later reached a speed of 262 miles per hour (422 km/h). The principle was considered for a high-speed link between Paris and Orléans, but government funding was withdrawn in favor of the *TGV*.

The development of the air-cushion principle for transport has since been overshadowed by work on magnetic levitation, although an experimental hovertrain running on solar and wind energy has been tested in Japan. It has four horizontal wings, and an unstaffed model

THE CYBERTRAN *test car (above) belongs to a Dual Mode Transport system. The artist's rendering (opposite) shows an ABT system. Both are under development in the United States.*

travels at 310 miles per hour (nearly 500 km/h) along a U-shaped channel lined with solar collector panels. The Wing-In-Ground-Effect—where the wings are pushed up by the downdraft of the craft, keeping it "in flight"—takes over at 62 mph (100 km/h). At speeds lower than this, the train runs on wheels.

MAGLEV

Magnetic levitation uses the repulsive and attractive forces of magnetism to lift and move the train. The vehicles have superconducting magnets on board and they run in U-shaped guideways that have electromagnets built into their walls. The electromagnets can rapidly change polarity, interacting with the magnets on the vehicle to provide lift, propulsion, and steering. Except for wheels that carry the train at speeds of less than 62 miles per hour (100 km/h), the trains have no moving parts and no motor on board.

Research into maglev transport started in the early 1960s. Trials continue in Germany and Japan, in particular, with projects actively being promoted elsewhere. In Germany the Transrapid consortium's test track has carried thousands of passengers at up to 260 miles per hour (420 km/h); in 1999, MLX01, the latest vehicle in the government-funded Japanese maglev project, reached a new record speed of 345 miles per hour (555 km/h).

At this stage in its development, the cost of building and operating a major maglev transportation system

MAGLEV TRAINS *use electromagnetic forces to suspend, guide, and propel train cars along a specially constructed guideway. Maglev trains can reach speeds as high as 345 miles per hour (555 km/h) because there is no contact between the guideway and the train—and thus no friction.*

will almost certainly be high. But its promoters claim it is the only system capable of traveling through densely populated areas at speeds of up to 125 miles per hour (200 km/h) with almost no disturbance to residents, and it may well provide future passenger services.

DUAL MODE TRANSPORT

Dual Mode Transport integrates the flexibility of personal automobiles and other vehicles with the efficiencies of an automated network of rails. Vehicles can operate independently on a normal road but are also designed to operate on a network of automated trackways. Once on the trackways, the driver can relax as the vehicle is automatically taken to its destination under computer control. The vehicles are still powered internally by a combustion engine or some other technology, but it is also possible to power the vehicles by electricity once they are connected to the rail network. Designed to operate many small passenger-carrying vehicles at speeds of up to 150 miles per hour (240 km/h), Cybertran is a Dual Mode Transport system under development in the United States.

AUTOMATED BEAMCARRIAGE TRANSPORT

Suspended high above the ground, Automated Beamcarriage Transport (ABT) systems operate large numbers of small, driverless vehicles in a computer-controlled network. The vehicles can be supported on top of the beams, as with a monorail, but there are technical advantages in suspending them underneath. Passengers and freight get on and off at elevated stations, or vehicles can be lowered from the beamway.

ABT systems replace rail and road networks and are of most use in urban and suburban areas. The systems have a mixture of vehicles, including small personal transport vehicles, people-movers for larger groups, and a variety of specialized freight vehicles. The ABT proto-type was SIPEM (Siemens Personal Mover), which is still running between two university campuses in Dortmund, Germany.

A second system has been built in Düsseldorf to connect airport terminals, hotels, and car parks with a railroad station, and other ABT systems are in development, including one in Seattle in the United States.

Marine Transportation

The future of ocean transport depends on designing ships that

are able to move people and freight faster.

Many of the great civilizations were built upon their ability to travel and trade across the seas in ships.

Ships are still the cheapest and most effective way of moving cargo across oceans, but passengers now prefer a more expensive, but much quicker, form of transport: air travel. In the future, however, ocean transport will become a lot faster, both through the redesign of ships and the use of new propulsion systems, so that one day it may even be able to compete with airplanes.

Ships have to battle their way through the water. As a ship pushes forward, water creates a wave called a "captive wave." The faster the ship goes, the more effort it takes to overcome the captive wave. Different shapes of ship create unique captive waves, which determine the vessel's theoretical top speed. Exceeding this limit takes a lot of expensive energy.

Broadly speaking, a ship's top speed is about equal to the square root of its length. Speed is also influenced by the load inside the ship and the volume of hull in the water (the wider the hull, the slower the vessel).

REDESIGNING THE HULL

There are three types of hull design that could yield faster ships. Multihulled vessels such as catamarans are one possibility. They have two

MODELED ON *the design of the Viking longship (left), the slender monohull provides a faster, but ultimately more unstable, ship.*

or more hulls joined by a deck and are about twice as efficient as an equivalent vessel with a single hull. They operate very well in sheltered, calm seas but not in rough water, when forces on the decks can snap the vessel in two. A bigger cargo increases the problem, since the weight pushes the vessel deeper into the water and slows it down still further.

A second hull of the future may be the slender monohull, or the "destroyer." The slender monohulls are narrow and light and use design principles first utilized by the Vikings. While the hulls can provide faster ships, the ships may be unstable because they are so narrow, and become even more unstable with heavy loads and when traveling at higher speeds. Even the top possible speed of a slender monohull would be too slow to use any of the newer, high-performance marine propulsion systems.

The third design solution is the semiplaning monohull, which is part of a marine transport idea called FastShip. These semiplaning monohulls have a deep V-shaped front end to slice through the water and a wide but shallow rear that curves upward under the water. As the vessel moves faster, the hull lifts it up out of the water, reducing drag. The FastShip hull becomes more stable as it gains momentum and can operate at speeds where new propulsion systems will work. FastShips operate well in all weather conditions, unlike slender monohull and multihulls. But while a FastShip hull can move at twice the speed of an equivalent monohull, it uses a huge amount of power.

THE SHAPE OF THE HULL *is often a compromise between the role of the ship and its speed. The hull of the USA (left), a specially designed two-rudder racing sailboat, takes shape in a Stockton shipyard in California. The boat was mounted as a potential American challenger for the 1986 America's Cup race, so the hull design is narrow, light, and built for speed.*

PROPELLING THE SHIP

Big propellers cannot push large ships above a speed of 30 knots. At this speed the pressure of the propeller makes the water boil, causing vibrations that can crack a hull. So to move big ships faster, propulsion needs to be reassessed.

Two possible driving forces are gas turbines and water jets. Gas turbines are used in small ferries. The latest gas turbine engines generate nearly the same power as similar-sized diesel engines while producing less than 5 percent of the polluting sulfur and nitrogen oxide gases.

Water jets use a water turbine powered by a high-speed engine and work on a similar principle to a jet aeroengine. Unlike conventional propeller-driven systems, the efficiency of the water jets increases with speed, making them ideal for high-speed vessels.

PROPELLERS *are still the most widespread propulsion device. Even so, a big propeller like the one belonging to the icebreaker SS Manhattan (above) couldn't push the ship above 30 knots.*

GOING SOLAR

Current solar technology cannot compete with conventional forms of energy when it comes to powering ships and boats. But a revolutionary concept undergoing trials in Australia may give solar-powered boats the push they need, with a little help from the wind.

The Solar Sailor has all its available surfaces covered in solar panels that collect power to drive an electric outboard motor. Many of the panels are arranged on moveable sails that can be raised and lowered to take advantage of any available wind, or angled toward the sun to optimize their light-catching potential.

143

LEARNING FROM NATURE

Fish don't use rotating propellers; they use fins that move from side to side. Although water is 800 times more dense than air, fish manage to reach speeds of 40 miles per hour (64 km/h) and can turn 180 degrees in their own length without slowing down. Most ships need about ten times their own length to U-turn and even then have to slow their speeds by half.

To better understand how fishtail propulsion works, engineers from the Massachusetts Institute of Technology (MIT) in the United States have built fish robots called RoboTuna. The two built so far are 4 feet (1.2 m) long. They propel themselves through the water by bending and flexing their "spine," like a real tuna. They swim in a tank, where they are attached to rails and controlled by an external computer.

Using RoboTuna, the MIT team discovered that fish create vortices that flow along the side of the body. Moving ships create the same vortices, but because the hull of a ship is inflexible, the hull resists the vortices, thus creating drag and slowing down the vessel. The more flexible spine of the fish allows these vortices to roll along the side of the body, and the tail then flicks against them, propelling the fish forward.

A team at Draper Laboratory in Cambridge, Massachusetts, in the United States, has used data from the RoboTuna to create an 8-foot (2.4-m) free-swimming Vorticity Control Unmanned Undersea Vehicle. It's used to investigate a variety of swimming strokes that are not available to the tethered RoboTuna.

The fish tail is an energy-efficient propulsion system ideal for small vessels, particularly for underwater research craft that need to carry all their fuel for extended voyages that may perhaps last for months. However, because the fish propulsion system relies on a flexible hull, it is unlikely to replace propellers in large, ocean-going vessels. But another MIT team is applying some of the RoboTuna results to conventional ship propulsion.

Successful trials have already been carried out on Proteus the Penguin Boat. This 13-foot (4-m) boat has two paddle-like foils instead of propellers, and these act like the flippers of turtles or penguins. They have already demonstrated 87 percent efficiency, which is around 17 percent more efficient than propellers, displacing a greater amount of water with each stroke than the blades of a propeller and giving a greater thrust for each movement. On a full-sized ship about 300 feet (100 m) long, the foils would need to flap once every two seconds to drive the vessel at 30 knots.

TAKING TO THE AIR

A variety of sea craft that doesn't get wet is also under development. The craft is essentially a small aircraft, but its wings are too small to lift the plane high above the ground. When the plane skims low over a flat surface, the wings are pushed up by the downdraft of the craft, keeping the vessel "in flight." This is known as the Wing-In-Surface-Effect (WISE), or sometimes the Wing-In-Ground-Effect.

THE SOLAR SAILOR *(below) is a revolutionary solar-powered boat undergoing trials. All available surfaces are covered in solar panels that collect power to drive an electric outboard motor.*

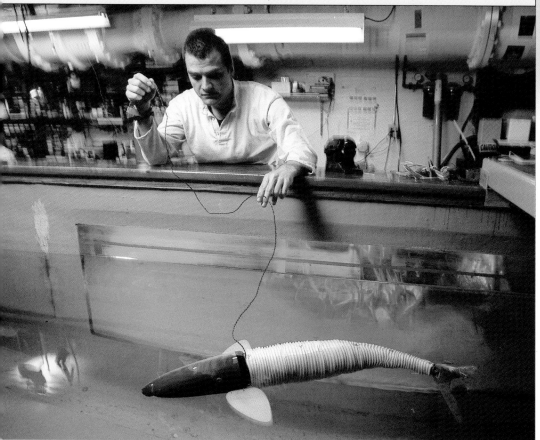

WISE craft are not really new. The first patent was granted in Finland in 1935, but the concept did not take off. The Soviet Union conducted a lot of development during the Cold War, and other countries including Australia, Japan, Germany, and Korea have also worked on the concept. The main drawback is that WISE craft can only operate over relatively flat surfaces, which requires a calm sea.

WISE craft are more energy-efficient than aircraft but otherwise behave in a similar way. WISE craft could be good for passenger or light freight work, promising the lowest cost–per-passenger-mile of all the existing and emerging transportation technologies. WISE ferries could operate at between 50 and 250 miles per hour (80 and 400 km/h) or faster, while carrying perhaps 25 passengers or more.

SUPERSONIC SUBS

A quantum leap in nautical transport may be at hand with supercavitation technology. Already used in advanced Russian torpedoes, supercavitation involves creating an air bubble around the submarine vessel by ripping the water apart faster than it can collapse back on itself. Since supercavitation vehicles have no part in contact with water, they have almost no drag and can travel extremely fast. In fact, supersonic speeds have already been achieved for supercavitating

THE ROBOPIKE (above), a robotic version of a chain pickerel, was designed to better understand how the flap of a thin fish tail can push a fish faster than a propeller can propel a ship. The Wingship (below) is a Wing-In-Surface-Effect vessel, which combines the best aspects of proven transportation technologies, namely catamaran, hovercraft, and surface-effect.

test vehicles, and theoretical speeds of 1.5 miles per second (2.5 km/s) have been proposed.

There are significant hurdles to the development of supercavitation technology. The supercavitation bubble is always quite small, and a vehicle needs to be completely contained inside it. A supercavitation vehicle also needs to be rocket-powered, but there are very few rocket engines small enough to fit within a supercavitation bubble. Finally, no one knows how to steer a supercavitating vehicle, since they were developed to be aimed and fired at a target.

Air Transportation

As we become increasingly reliant on air transport, the search is

on for new ways to carry more people through the skies—and

to do it more quickly and cheaply than ever before.

Air travel is one of the most important forms of transport today. It is quick and comfortable, but expensive. By 2020 the world's cargo airplane fleet will have doubled in size to almost 3,200 aircraft and there will be about 32,000 passenger airliners in operation around the world. Most new passenger aircraft will be modified versions of existing models, like the McConnell-Douglas DC-10, Boeing 747, 767, and Airbus, but larger planes that take more people are also in development.

The new Airbus A3XX-100, seating 555 passengers, will enter service in 2005. It will be the first passenger aircraft with two decks running the length of the plane. Later models are expected to carry about 650 people, half as many again as the current Boeing 747s.

THE FLYING WING
Not to be outdone, the Boeing research team is looking at a radical design that will give a greater payload and reduce running costs. The Blended Wing Body (BWB) design replaces the cylindrical fuselage and attached wings with a flying wing, a concept first floated in the 1940s. The advantage of this design is that the whole aircraft generates lift, but this increases drag, necessitating more-powerful engines.

Now that lightweight, powerful engines are available, the BWB is a commercially viable design. Variants that could carry 250, 450, and 600 passengers are being considered, together with proposals for 170-seat and 800-seat aircraft. If all goes according to plan, commercial BWBs could replace existing larger airliners by 2010, with military versions becoming available a number of years beforehand.

GOING FASTER
Getting planes to fly at ever-greater speeds raises some problems, the main one being air density. The faster a plane travels, the more it

MORE SPEED *and space are the demands being placed on designers of new aircrafts around the world. The ramjet engine (above, left) may be replaced by "scramjets," which could meet the first demand. The second is glimpsed in this full-scale model (above) of the forward section of the A3XX-100 double-decker passenger jet, being moved to a new hangar. When in service in 2005, it will be the largest aircraft in the world.*

compresses the air in front of it and the more difficult the air is to move through. Researchers, using ideas developed by Russian engineers during the Cold War, are bombarding the air ahead of a plane with a beam of microwaves. This ionizes the air, reduces drag and allows the plane to slip through more easily. There are still technical difficulties to be overcome, but this concept promises faster speeds, lower fuel consumption, and some stealth abilities, too.

There is also the problem of engine capabilities. Current ramjets need to slow incoming air from a supersonic speed to a very low velocity so the air can ignite in the combustion chamber and provide thrust. This need to slow incoming air acts as an effective top speed at which engines can operate. A possible solution to the problem could be the scramjet, a new engine that still has to slow incoming air for combustion, but not to subsonic speeds. This could be the engine that powers the first plane to travel at Mach 4 (four times the speed of sound) or faster. Such an aircraft would need to have a

separate engine for low-speed flight because the scramjet can only operate at higher speeds.

TAKE-OFF

Increasing air traffic will place greater demands on the already-strained airport facilities around the world. In addition, larger planes will need longer runways and larger airports—a problem when existing airports are often located close to cities, where land is expensive. One solution could be to get planes into the air quicker, on shorter runways. (A better method of launching planes could also result in significant reductions in fuel consumption, because take-off is the most fuel-hungry part of a flight.)

The United States and British navies are considering using electromagnetic catapults to replace the steam-powered catapults currently used to assist take-offs from aircraft carriers. These catapults use a series of magnetic fields to rapidly accelerate a plane to take-off speed. If successful, the electromagnetic catapult could be used for civilian aircraft. The drawback, however, is that most people would probably be uncomfortable with the gravitational forces experienced during such rapid acceleration.

SMART PLANES

The design of any plane is a compromise. This is because a plane's optimal theoretical shape alters subtly with the slightest change in speed.

147

The perfect plane would change its shape as it changes speed—the concept of the smart plane.

Wing surfaces in particular need constant trimming and reshaping for optimal flight performance. The wings of a smart plane would be covered in microscopic mobile panels. The panels would be jiggled in flight by actuators responding to thousands of tiny sensors linked to a network of computers around the aircraft. Turbulence would be better controlled by this continual reshaping of wing surfaces, leading to a more economical and smoother flight. Such smart planes, though theoretically possible, are still some way off production.

THE AIRSHIP

The oldest proven aircraft technology is the airship, and it has a bright future—using non-flammable helium as a safe alternative to the highly flammable hydrogen that was used in the 1930s. Airships use less than 10 percent of the fuel consumed by a comparable airplane. They are stable in the air, they can hover for hours, and their low vibration and noise make them excellent platforms for sensitive equipment and powerful sensors.

Airships are currently being developed for maritime and military surveillance, geological and biological survey work, and meteorological and air pollution monitoring. They also have advertising, mapping, and law enforcement uses. Passenger airships are being explored, too. Although slower than fixed-wing aircraft, airships offer about twice the space per passenger, they do not need airports (passengers could be

INCREASING AIR TRAFFIC AND LARGER PLANES *will mean airports will require more and more space for landing and take-off. But land is expensive. One way to limit the length of runways would be for passenger jets to take off using a catapult like those used on aircraft carriers (above).*

picked up and put down in the middle of a city), and they can travel directly from point to point, giving faster journey times than high-speed trains. A passenger airship could complete short-haul intercity services in roughly twice the time taken by a commercial flight but for less than half the cost.

MILITARY AIRCRAFT

The F-22 Raptor, under development in the United States, represents the next generation of fighter aircraft. The F-22 will enter service in 2005 at an estimated cost of $83.6 million each, and production will run through to 2013.

Capable of Mach 2 or faster, the F-22 has unprecedented maneuverability while being able to operate at slower speeds than most other fighter aircraft. Stealth technologies give the F-22 a radar signature similar to that of a bumblebee. This is achieved by reducing radar-reflecting surfaces (all weapons are carried internally) and using radar-absorbing surfaces.

Designed to work in tandem with the F-22, the Joint Strike Fighter (JSF) will be a fighter aircraft that can also act as a strike-bomber, a fast plane that handles well at low speeds and can make short take-offs and vertical landings. Production of the JSF has not yet begun, and serious questions have been raised about the wisdom of developing this expensive fighter aircraft when there is no indication that a rival nation is developing a similar weapon.

BEYOND THE SKIES

NASA has been trying to develop a Blended Wing Body design that integrates the cabin into the wing, a design akin to the stealth bomber.

NASA would like to incorporate this concept into a reusable suborbital vehicle. Such an aircraft would launch vertically into a low-Earth orbit and land like a conventional plane. It could allow travel times of just 90 minutes between London, England and Sydney, Australia. This concept has been around for a long time but has proven difficult to achieve.

The X-33, a scale model of the proposed prototype of a Reusable Launch Vehicle (RLV), was showing promise as a viable suborbital vehicle design concept. Powered by two revolutionary linear aerospike rocket engines and shaped like a flying wedge, development of the X-33 ceased in early 2001 due to budget cutbacks in the United States by Congress.

An RLV is intended as a replacement for the space shuttle to carry satellites into orbit. If successful, the RLV should reduce the cost of getting a payload into space by 90 percent.

THIS F-22 *Raptor is being tested over Edwards Air Force Base, California. It is set to succeed the F-15 Eagle. The F-22 features low-radar observability, supersonic cruise, six ASIM-120C missiles, two AIM-9 missiles, and a 20mm Cannon.*

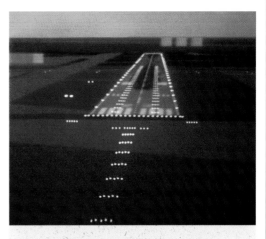

ROOM WITH NO VIEW

Cockpit windows are slowing the development of supersonic planes. The pilot needs to see out of the aircraft but windows create aerodynamic drag. The High Speed Civil Transport program at NASA is replacing cockpit windows with an External Visibility System (XVS). One step up from the flight simulator above, cameras and other sensors will relay information to the pilot inside a windowless cockpit. Apart from possessing increased speed capabilities, aircraft fitted with XVS could take off and land in conditions that would ground ordinary aircraft.

Development of an RLV would solve many of the technical hurdles that still stand in the way of a commercially viable sub-orbital airliner. However, budget restrictions against the funding of the development of an RLV have put the project on hold indefinitely.

The Future of Space Travel

If flying in space is ever going to become commonplace,

it is time to develop some new technologies.

For sending tons (tonnes) of material from Earth into space, the acceleration and sheer lifting power of a rocket can't be beaten. Since the 1940s, chemically fueled rockets have launched thousands of payloads, some weighing many tons (tonnes). And big rockets still reign supreme for sending large cargoes into low-Earth orbit, or for sending smaller payloads to the planets.

A brute-force approach isn't always ideal once a payload reaches space, however, but until now mission planners have had little else to choose from. That is all about to change, as several new technologies come into view on the near horizon, promising innovative and more efficient ways of moving payloads.

EASY DOES IT

Picking up cargo from ground level and putting it into orbit will remain the province of large, powerful rockets for decades to come—

the need to battle Earth's gravity virtually requires them. But once a payload is in orbit around Earth, sending it somewhere else—to Mars, say, or to a comet—does not need the high acceleration and high thrust of a chemical rocket. An engine producing a much lower thrust can do the job just as well and can do it more efficiently. But there is a catch: Because such an engine provides only very gentle accelerations, travel times may be long.

This "new" technology has actually been tested on the ground since the early 1960s. It is called the ion drive, or solar electric propulsion. It works by giving atoms (such as xenon) an electrical charge and accelerating them in an electric field to speeds of about 20 miles per second (30 km/sec). The stream of ionized atoms shoots off the back of the engine, producing a very small thrust, about equal to the force exerted by a sheet of writing paper on your hand.

It hardly sounds like the gateway to the planets, but an ion drive can run for months at full throttle, building up some very respectable velocities. And ion drives work: In October 1998, NASA launched Deep Space 1, a technology-testing mission on a (chemical) Delta II rocket. Once it reached escape velocity from Earth, Deep Space 1 fired its ion drive. This powered the spacecraft to a flyby with asteroid Braille in July 1999 and then on to another flyby with Comet Borrelly in September 2001.

Deep Space 1 is merely a test; engineers expect to build much more powerful ion drives soon. The thrust depends only on the strength of the field that accel-

WHEN COLUMBIA, *the first space shuttle, launched on April 12, 1981 (left), it revolutionized space travel. Today there are four space shuttles and, despite their age, it may be some time before we see a new design.*

ONE ENGINE *being tested for interplanetary spacecraft is the ion drive. The plasma contractor (above) allows cathodes to neutralize the negative charge induced by the engine. An artist's concept of NASA's Deep Space 1 (right) depicts it passing asteroid Braille in July 1999, propelled by an ion drive.*

erates the atoms. This is ultimately determined by the amount of electric power available. Since sunlight is most plentiful inside the orbit of Jupiter, ion drives may soon be buzzing around the inner solar system.

SAILING THE LIGHT FANTASTIC

Ion drives aren't the only low-thrust power-plants being worked on. Another idea is solar sailing—using the pressure of sunlight to blow a spacecraft from one planet to the next.

Such a spacecraft would look like no other craft that has ever flown. It would have a huge, lightweight sail, which would be perhaps half a mile (a kilometer) on one side and highly reflective. Rigged on lightweight, composite-material spars and controlled by computers, the sail would unfurl to catch sunlight. (Don't confuse sunlight with the solar wind. The latter is a flow of charged particles that also comes from the Sun, but which would exert only a negligible force on a solar sail.)

At Earth's distance from the Sun, sunlight would press on every square yard (meter) of sail with about the weight of a small coin. The push would be gentle, but would never stop. The spacecraft would be steered much like a sailboat is on Earth—to change the direction of acceleration, you would simply need to tilt the sail at an angle to the sunlight.

Before anyone hoists sail for the planets, however, some daunting engineering hurdles need to be overcome. A small-scale experiment

151

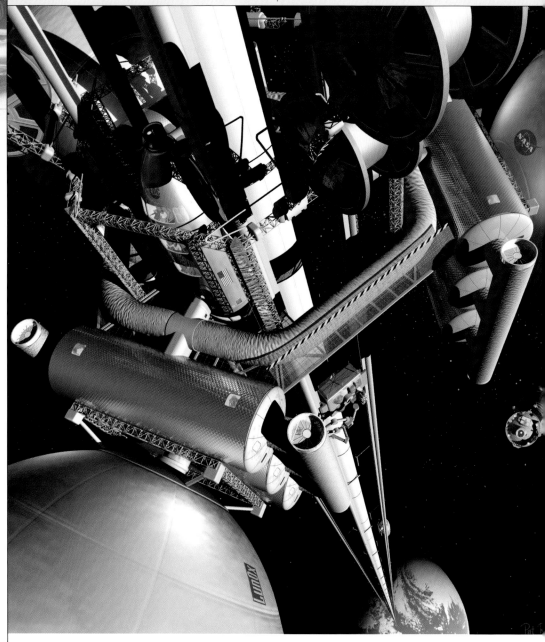

SPACE TRANSPORT *could become more accessible toward the end of the 21st century if the space elevator is built. This drawing shows how it would connect Earth to a geostationary Earth orbit, its center over a base fixed at the equator.*

with a lightweight reflector was performed successfully on a Russian spaceflight in 1993. But it is not clear how to manufacture, package, and deploy in space gigantic sheets of very thin aluminized plastic. And then there's the problem of how to control them. But there are no theoretical barriers to the development of solar sails, and the 21st century could see fleets of sailing spacecraft tacking across the planetary realm in much the same way as their ancient forebears did on the oceans of Earth.

A TOWERING ELEVATOR

If solar sails sound weird, downright bizarre is the idea of the "space elevator." The concept is a cable running from Earth's surface to geostationary orbit, a point at an altitude of 22,200 miles (35,800 km). The plan is that cargoes

could be shuttled along the cable as easily as elevators in a department store. No roar and thunder of rockets, just the whoosh of electromagnetically driven vehicles going up and down a cosmic beanstalk into space. Once lifted to geostationary orbit, spacecraft could leave for destinations in the solar system using conventional rockets or other means.

The physics is impeccable, but even advocates admit that the engineering would be very tough. The elevator would have to be built by sending a spacecraft to geostationary orbit over the equator. Within a period of 24 hours, the

satellite would remain fixed in the sky over one spot on Earth. From there, the satellite would drop a cable to the ground, and extend a cable of equal length upward from its orbit for balance. A tower 30 miles (50 km) high would then anchor the bottom of the cable.

The idea of a space elevator was popularized by Sir Arthur C. Clarke's 1978 novel *The Fountains of Paradise,* but the idea has been independently stumbled upon several times since the 1890s. In 1999, a NASA workshop studied the idea and developed plans for its implementation. These plans identified the necessary technologies, which include high-strength materials for the cable and for the tether tower, the ability to deploy and control such long structures, and electromagnetic propulsion systems for the vehicles.

Clarke has said that the space elevator would be built "probably about 50 years after everybody quits laughing." But the NASA workshop concluded that such a project would be feasible toward the end of the 21st century.

GRAB AND FLING

The space elevator's vehicles glide up and down the cable using electromagnetic propulsion such as a linear electric motor. But imagine that such a motor were used to accelerate vehicles to spacefaring speeds. The result would be a mass driver—an electric catapult that passes a cargo container along a rail assembly at a steadily accelerating rate until the container flies off the end at escape velocity.

Such a device could not work on a planet with an atmosphere, thanks to air friction, and it won't work drifting in interplanetary space. The mass driver has sufficient recoil that it needs to be anchored to a body massive enough to be unaffected by the launch forces. But a mass driver would work on the airless Moon or a large asteroid, where it could fire containers of processed minerals or lunar ore into space for pickup by cargo craft.

Mass drivers were developed starting in the late 1970s by Gerard K. O'Neill. Lab-scale versions have been built, and they demonstrate that the principle works perfectly. In essence, this is a technology that's ready to be scaled up to working size. The chief thing holding it back is the lack of an established lunar base or a mining operation in the asteroid belt.

GIVE NUKES A CHANCE

A mainstay of old-time science fiction was the nuclear rocket engine, which delivered the power to take heroes like Space Cadet Tom Corbett across the solar system in time to catch the villain. But nuclear rockets weren't just fixtures of sci-fi space operas. Real nuclear rocket engines like the one above, which was to be part of a mission to Mars, were test-fired in Nevada during the 1960s. The program, called NERVA (Nuclear Engine for Rocket Vehicle Applications) had a nuclear reactor that heated liquid hydrogen and shot it out of a rocket nozzle. Project NERVA was shut down in 1972 after support for most nuclear technology development evaporated in the United States.

The idea, however, hasn't been completely forgotten, because it offers a lot of power compared with chemical rockets. NASA engineers have drawn up plans for a nuclear-powered rocket to carry crews to Mars in less time than chemical or other kinds of propulsion would get them there. This has great benefits, as astronauts would be exposed to zero gravity for a shorter time, with fewer biomedical problems.

A nuclear Mars rocket would not lift off from the ground. Chemical engines would get the rocket into orbit, from where it would leave for Mars. The trip out might last just four months, half the time for a chemically fueled trip. The crew could spend a month or so on Mars, and the return would take eight months. They could be back after just over a year. And refueled with liquid hydrogen, the craft would be reusable.

It's unclear whether the general public would support the development of a nuclear rocket, even though the technology is now better understood and the benefits for spaceflight are clear cut. Yet even without a nuclear option, there are many developments in the works to make spaceflight more efficient, less expensive, and far more widespread.

CHAPTER FIVE

INFOTECH

I think there is a world market for about five computers.

Attributed to THOMAS J. WATSON (1874–1956), American industrialist

The Future of Computing

The days of the silicon chip are numbered. In its place could come machines built from spinach, DNA, or single atoms.

olossus, the world's first programmable electronic computer, whirred into life in England in 1943. Designed to crack German codes, it was a behemoth that almost filled an entire room. Two years later, United States scientists threw the switch on their computer ENIAC. This machine was a monster the size of a small fleet of English double-decker buses. One day, speculated *Popular Mechanics* magazine in 1949, computers might shrink to the size of an automobile.

Half a century later, the future of computing is every bit as difficult to forecast. Who would have predicted the rise of e-mail, for instance, or guessed that you would find computer processors in everything from bicycles to birthday cards?

Yet signs of things to come are beginning to emerge. There are prototypes for computers that run on nothing more than light beams. Researchers have taken the first steps toward building computers from single molecules, and from proteins extracted from plant leaves. The first calculations have been made using a computer built solely from DNA. Even the weird world of quantum mechanics has its star—a quantum computer that will one day wield unimaginable power, and yet it may not even need to be switched on.

Right now there seems to be just one thing that's certain: The days of the silicon chip are numbered. In 1965, a computer engineer by the name of Gordon Moore suggested that the size of the circuits etched into each silicon chip would halve every 18–24 months. "Moore's law" has held up remarkably well ever since. However, all kinds of indicators—including the laws of physics—suggest that we can't go on shrinking silicon circuits forever.

ON A RAY OF LIGHT

Finding new ways to miniaturize these already tiny circuits is hard enough. And once a silicon processor reaches a certain size, it becomes virtually impossible to supply it with all the information—control signals, data, and the like—

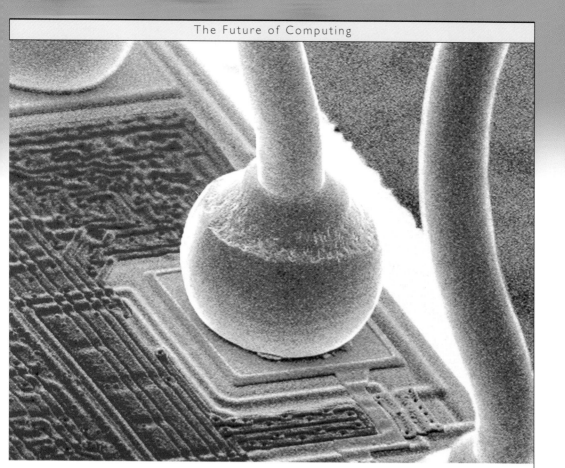

COMPUTERS BASED ON *silicon chips (above) could soon be heading for a high-tech graveyard much like this one (above, left) in Japan. The problem is that size matters. As chips shrink ever smaller, their metal wires take up more available space, squeezing out other vital components such as transistors. Eventually it will become almost impossible to supply the processors with all the signals and data they need to work efficiently.*

that it needs to run efficiently. Even today, about 70 percent of the space in a processor chip is wiring. But what if we could get rid of metal wires altogether and replace them with light beams? Light beams can be thinner than any wire, and you can run beams through each other without their signals interfering.

This kind of hybrid computer is already in development. The machine combines powerful silicon-based processors with an array of lenses and mirrors that bounce laser light from one chip to another. Compared with metal wires, these light beams are data "fire hoses," capable of squirting thousands of times more digital information every second from one spot to another than conventional connections. Best of all, light "wires" take up far less space than their metal equivalents, so computer engineers can pack far more into the same space.

Laser beams may give silicon a new lease of life, but eventually the laws of physics will intervene to finish those tiny chips off. At some point in the future—by about 2020 according

to physicist Neil Gershenfeld from the Massachusetts Institute of Technology—silicon processors will become so small that the uncertainties of quantum mechanics will step in. This means that when the dimensions of transistors reach a few tens of nanometers (billionths of a meter), the electrons they carry will no longer be confined inside. They'll leap from one transistor to another and the silicon computer as we know it will short circuit and die.

Yet, according to Moore, there is another factor that will kill silicon long before quantum mechanics can come into play: economics. Moore's second law predicts that the cost of retooling the silicon chip industry is rising so fast that it will soon outstrip the profits to be made from the chips themselves. Halve the size of its circuits and a chip costs five times more to make. By about 2010, the cost of a single manufacturing plant could reach $100 billion, a significant proportion of the annual chip market. Suddenly it will no longer make sense to stay with silicon. When that time comes, what computer manufacturers will need is a new breed of computer processor built from cheap components that are easy to assemble.

MOLECULAR SOLUTIONS
American chemists such as Jim Tour from Rice University in Houston, Texas, and Mark Reed at Yale University in Connecticut believe they

157

have found the answer. Tour, Reed, and many others are in the process of creating the building blocks of computer electronics—the logic gates, wires, and transistors—from small assemblies of single molecules.

The first molecular components to appear were simple wires based on polyphenylene, a molecule made from a short chain of carbon rings. Then came the molecular diode, a switch that lets current flow in one direction only. Diodes are pretty useful. Link enough together and you can create simple logic gates—the building blocks of transistors.

Recently chemists created the molecular equivalent of a memory chip, "atomic solder" to join the parts together, and even insulation that will stop molecular wires short-circuiting. With most of the components in place, we could soon see the first prototypes of molecular processors so small that millions of them could fit on the end of a pin.

GREEN COMPUTERS
Other chemists are borrowing components from nature. They've discovered that tiny proteins extracted from the leaves of green plants can be connected with nanowires to make electrical valves much like diodes.

ONE OF THE FIRST quantum computers has been developed by Neil Gershenfeld (above) at the Massachusetts Institute of Technology. The device is based on a vial of water surrounded by magnets. Computers like this could soon replace silicon-based computers if Moore's second law holds true. This law (illustrated right) predicts that fabrication plants for silicon chips will soon cost a large fraction of the total chip market.

These devices could even be powered with sunlight. Some plant proteins contain chlorophyll, the molecule that colors leaves green. Chlorophyll absorbs sunlight, which is used to power photosynthesis, so you could connect arrays of these light absorbers to a processor built from plant proteins, and have a solar-powered chip that's green in every sense of the word. If plant-based electronic circuits do hit it big, they could even herald a new age of computing that sees the factories of Silicon Valley replaced by vast fields of spinach or pea plants.

Whatever form it takes, molecular computing is without doubt a force to be reckoned with. The components are cheap and thousands of times smaller than silicon transistors. Just as chemists throw chemicals together to trigger a reaction, one day computer engineers could mix all the molecular components they need for their circuits and produce molecular computers in a test tube.

This kind of molecular "engineering" is rarely perfect in that not every molecular component they make will be up to scratch. Yet surprisingly, it may not even matter. You don't need perfect chips to create powerful machines. To prove this, engineers at Hewlett Packard in California have constructed a remarkable new computer called Teramac.

Teramac is the first of a new breed of fault-tolerant computers built from both perfect and imperfect silicon chips. In fact about three percent of its hardware is defective, yet Teramac is programmed to spot the faults and route data around them. Warts and all, this machine is hundreds of times more powerful than a top of the range workstation.

The same idea could easily work on a much smaller scale, too. Chemists have already used a similar approach to construct computer processors from thin tubes of carbon atoms called nanotubes. Particular kinds of flaws in nano-

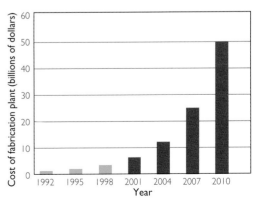

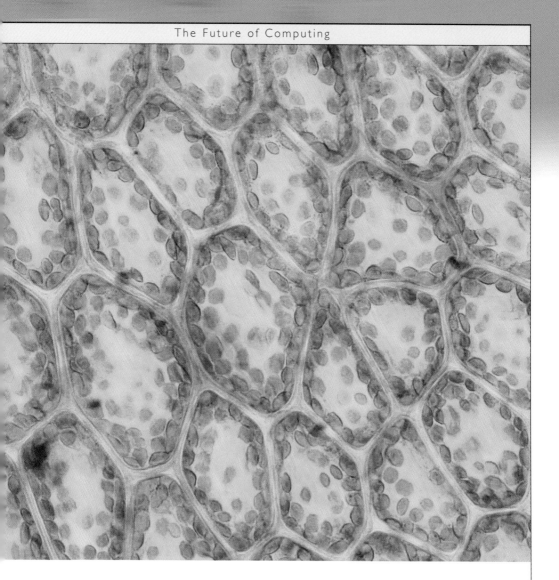

GREEN LEAVES *could offer a radical way to build the computers of the future. Tiny proteins and green chlorophyll—as seen in this light micrograph showing a section through a leaf—can be used to manufacture transistors and other electronic components when extracted from plants.*

tubes can turn them into diodes or even transistors. So if you create a block containing trillions of nanotubes, you'd expect to find arrangements of "faulty" nanotubes that together can form useful electronic circuits. Finding them means testing the cube at many points, but once you've identified enough components, you can connect them into useful circuits. The same technique might one day be used to build immensely powerful, fault-tolerant, molecular computers.

GENE MACHINES

A more radical plan is to abandon traditional notions of digital computing altogether. Take DNA, for example. It stores a huge number of programs as combinations of four bases that are arranged along its backbone—the rungs in the double-helix ladder. Molecules called

ribosomes, which read the code on the DNA and use it to manufacture proteins, run these programs. Is there a way to use programs stored on DNA to mimic traditional computers?

Almost certainly. Mathematicians have known for years that if you take small tiles like dominoes and code them with data and simple rules that govern the way they fit together, you can use them to add or to multiply numbers. If it works with tiles, why shouldn't it work with small blocks of DNA just 15 billionths of a meter across?

Leonard Adleman, a computer scientist at the University of Southern California in Los Angeles, has already shown that this is possible. He used small blocks of DNA to represent parts of a mathematical problem and then mixed billions of strands in a test tube. The DNA immediately bound together according to the instructions on their surface, creating a mixture that contained every combination of solutions to the problem. In this case Adleman wanted the simplest solution so he carefully separated the shortest combination of DNA sequences and read the answer off.

159

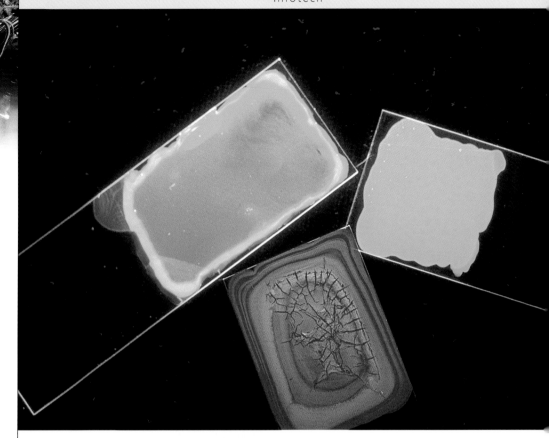

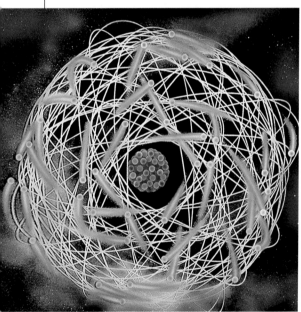

FOR THE SMALLEST *of computers, physicists are turning to individual molecules and atoms (left) to construct components. Molecules like polyphenylene vinylene (above) conduct electricity, and when connected together into tiny circuits act like miniature wires. Other molecules can play the roles of transistors and switches, allowing complete molecular circuits to be constructed.*

ULTIMATE CALCULATOR

More advanced, and potentially far more powerful, is a quantum computer. Rather than working with "clunky" technology like diodes, the quantum computer relies on fundamental quantum particles such as atoms or electrons. But it's not the size of its components that makes quantum computing so powerful, it's the bizarre laws of quantum mechanics it uses.

In the familiar world of digital computers, information is represented as bits of data that can be either 1 or 0. But in quantum computing, a single bit—termed a qubit—can exist as both 1 and 0 at the same time.

Take the spin of an electron, for example. It can spin in two directions—a clockwise spin representing 1 and an anti-clockwise spin that represents 0. But an electron can also exist in a weird quantum mechanical state called a superposition, where it spins both ways at once. Use an electron to make a calculation while it is in this state and it will do the job for both the 1 and the 0 simultaneously—two calculations for the price of one.

At the moment, even advocates of DNA computing admit that it probably won't rival silicon—at least not for another 50 years or so. The problem is building the DNA blocks and reading the result once the computation is finished. It isn't impossible—it just takes a lot of lab work and careful analysis. But this simple biological computer could push technology in some unexpected directions. Eventually it might even be possible to program living cells to make calculations.

COMPUTING WITH LIGHT

Back in the 1960s, the concept of an all-optical computer emerged. Rather than electrons, an optical computer trades in photons. Processing takes place in a slab of special glass, data is stored in some kind of holographic memory device, and information is passed between the two on laser beams. This machine would be unimaginably fast, but building one has so far proved an impossible challenge. For the moment it seems that optical switches and optical memory are simply too complex to work reliably, but that could change in a decade or so.

Better still, the more qubits you add to your quantum computer, the more powerful it becomes. With two qubits you can compute four numbers at the same time. With a few hundred qubits, your computer could simultaneously work on more numbers than there are atoms in the universe.

So what does a quantum computer look like? So far physicists have constructed some simple computers that contain just a few atoms, cooled to around 4° Kelvin (-452°F/-269°C). To do a calculation, they control the state of each atom with a burst of radio waves. When an atom absorbs the radio waves, it flicks from a 0 into a 1, for example.

To get at the results physicists will have to learn to use a quantum algorithm—the quantum equivalent of a computer program—and the correct solution should literally "fall out." The secret is a phenomenon called quantum interference. Physicists must design their quantum algorithms so that the undesired answers interfere with each other destructively and disappear like out-of-phase light waves. This way, the desired answer is the only one that remains. In fact, some scientists believe that the strange rules of quantum mechanics may one day enable computer scientists to get at their results without even having to switch their creations on.

Quantum computing is still in its infancy. Just as with Colossus and ENIAC in the 1940s, the first users of quantum computers will probably be scientists in military labs. The sheer power of these quantum machines makes them perfect code breakers. But no doubt academic researchers will be hot on their heels. If these machines live up to their promise, quantum computers could open the way to the quantum Internet where qubits would be traded like currency. They could solve complex problems way beyond the scope of any traditional computer. They could even probe the mysteries of quantum mechanics itself. You might perhaps have one on your desktop one day. The truth is, we can't begin to imagine where the power of these machines might lead.

NATURE INSPIRES *exciting proposals for computers of the future. The human nervous system is a remarkable model of complex circuitry, and one day it might be possible to link DNA or living cells to integrated chips, as illustrated in this artwork..*

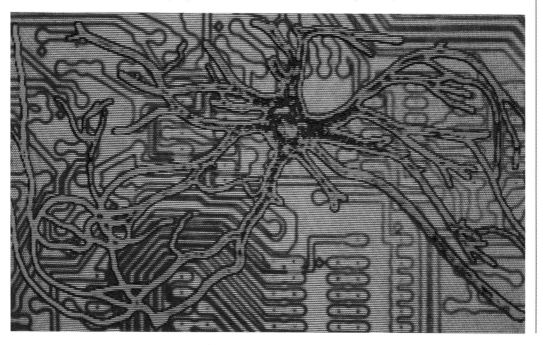

Wireless Networks

Combine digital electronics with wireless communication

and modern location systems, and you can

instruct the microwave how to cook your meal while

driving home, and jog in clothes that spur you on.

As you pull into your driveway the front door opens automatically at a signal from your mobile phone handset. You are greeted by a repairman who has just arrived in response to a message that was sent by your washing machine. It needs a bearing changed. The washing machine isn't the only appliance that has been on the airwaves. As you passed the local shopping center, the fridge reminded you to bring home some milk. While putting away the milk you notice there is a video message from your daughter on the touchscreen on the fridge door. She's gone to tennis without you. You use the touchscreen to switch the television on and then settle down to watch it.

Welcome to the era of Bluetooth, a computer chip that allows wireless communication between all the electronic devices in your home—and, via the Internet, with the world outside. Bluetooth, named by Swedish telecommunications company Ericsson after a Danish king who enforced order on the unruly Vikings, is part of a wireless revolution which will change societies all over the world. By

2006, investment bank Merrill Lynch estimates, people will be collectively buying two billion Bluetooth-equipped devices a year.

The Bluetooth chip incorporates a low-power, short-range radio transceiver that can talk to all the other Bluetooth chips within a 33-foot (10-m) range. Inside the house, this means that all electronic devices could potentially be controlled from a central console, most likely in the space now occupied by the Post-It notes stuck on the fridge.

But incorporate Bluetooth into a mobile phone and you can take your controller on the road, communicating directly, or over the telephone network or the Internet, with any Bluetooth device that will allow you. It would be possible, for instance, to transfer money from your bank account to pay for a purchase in a store anywhere in the world, to program the watering system in the garden of your vacation home 600 miles (960 km) away, or simply to switch off your office light. And all these things can be done from wherever is convenient—a car, a bar, or a sidewalk.

But that's not all your mobile phone will allow you to do. Already, via Wireless Access Protocol (WAP) and several other systems, you can surf an expanding wireless version of the World Wide Web. Soon, when the speed of wireless data transmission increases sufficiently, you will be able to watch television or listen to a CD on demand.

INFORMATION ON THE MOVE
So, mobile phones are evolving into universal communicators. One of the driving forces behind this evolution is the convergence of all forms of information provision—radio, television, telephone, and the Internet. Digital

WITH THE ADVENT OF *wireless communications and a scanner to check off items as they are removed, your fridge will be able to call the supermarket with a shopping list for home delivery when food and drink begins to run low.*

electronics reduces all forms of information—color, sound, video, print—to numbers. Information can then be analyzed, processed, mixed, matched, and transported in the same way. And the most efficient form of transferring such digital information is via packet switching.

Packet switching involves chopping up a piece of information into fragments or packets, like the picture in a jigsaw. Each packet is then carefully labeled as to its destination and where it fits into the whole jigsaw. The process is rather like disassembling a building, numbering the pieces, transporting them and then reconstructing the building in another location.

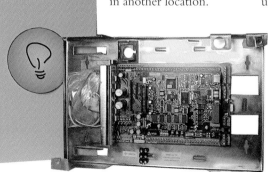

RAILWAY LINES *snaked their way across the landscape of industrial nations in the 19th century; in the 20th century it was telephone and power lines. But in the 21st century it will be communications antennae (above) which will become the symbol of development across the globe.*

Several different modes of electronic transfer have been devised and implemented. Some of these varying modes send each individual packet of information to its destination via whatever route happens to be open at the time. Packets can arrive in any order depending on how far each one has traveled, and are then ordered to form the whole message. This is fine for graphical information and makes efficient use of the communications networks. Other modes of transfer take the time and effort to establish a temporary connection through to the destination first, and then send all packets via that route, in order. Although

PROGRAM YOUR VCR, *water the garden, or turn down the thermostat while away on vacation—already it can all be done by calling this Thinkboxx controller (left) from your mobile phone, and pressing a few buttons.*

163

LIKE PIECES OF A JIGSAW, *in packet switching information is split into chunks or packets, each of which can be independently transmitted to a destination. But the message can only be reassembled in one way to make sense.*

these modes demand more time initially to organize, they ensure better transmission of sound and video, which can suffer somewhat in quality if critical packets are delayed.

WEARING THE FUTURE

Portable communication devices must be carried somehow. Almost as soon as mobile phones emerged on to the market, so did the pouches and belts to transport them comfortably. But as universal communicators develop, the link with clothing will become ever closer. Wearable computers are already available; the processor and circuitry fit into a pocket, the battery is worn on a belt, the keyboard wraps around the wrist, and the tiny computer screen projects from a pair of glasses. The Levi-Strauss company already sells jackets wired with connection points for the wearer to plug in his or

CELL PHONE TECHNOLOGY *can now reach remote areas much more quickly and cheaply than conventional cables. Countries in the developing world may well use wireless communications to leap-frog into the 21st century.*

her mobile phone, MP3 player and other electronic gear.

But what is planned is much more ambitious—smart clothing with the ability to generate its own energy from body heat and movement in order to power sensors and any other electronic equipment which will allow it to respond to your every need. On the drawing board is clothing which changes color with your mood or the weather, which releases aromas and pheromones at appropriate times, and which measures and analyzes your progress while you jog and whispers encouraging messages through your earphones.

Many of the proposed applications for the wireless world require a knowledge of the location of the user. On the more serious side is the demand by the United States Federal Communications Commission that future mobile phone handsets be able to be tracked once an emergency call is logged. But there are also plans for localized information, advertising, and services to be delivered to passers-by through their communicators. Such accurate location of mobile communicators is possible with the GPS system (see Box), but this raises questions of privacy. In future, those who wish to make a private visit, should perhaps do so with their mobile phones switched off.

YOUR PHONE NUMBER IS YOU

But even that may not be possible with the advent of the personal communications system. Here, as long as the system knows where you are, all calls, information, or services would automatically be routed to the nearest communicator. The idea is that the mobile phone becomes less important than your personal communication number (PCN). The PCN will allow you to place and receive calls, transact business and receive services anywhere in the world, using whatever device is handy. Finland, where much of the research into the wireless future is being undertaken, is even trialing the concept of using the PCN, or something very similar, as a standard, official piece of identification—the role which more often than not is filled by drivers license or passport numbers at present.

The impact of wireless communications may be even more profound in developing nations than in the industrial economies. It may allow

WHERE AM I?

The United States Federal Communications Commission (FCC) has specified that emergency services in the United States must be able to pinpoint anyone who calls in on a mobile phone. This can be done most simply by incorporating a cheap computer chip into the handset which allows the phone to communicate with the satellites of the Global Positioning System (GPS).

Originally developed for the United States military in the 1970s, GPS has been fully operational since 1994, although the system was used even earlier than this—in 1991—to guide the allied forces through the trackless deserts of Iraq during the Gulf War. While the technology seems staggering, the principle is really very simple.

A set of 24 satellites orbits the Earth at a distance of about 20,000 kilometers (11,000 nautical miles). Each of them continuously broadcasts signals containing the exact time and its precise position in space. Even though these signals travel to Earth at the speed of light, there is a detectable time delay between the time a satellite signal is broadcast, and the time it is received on Earth. This time delay can be used to calculate the distance the signal has traveled—that is, the distance from the receiver to the satellite. Receivers at that exact distance can only be somewhere along an arc across the Earth's surface.

But, at any one time, a GPS receiver should be able to detect signals from between four and six satellites. When those distances to the satellites are combined with information on their position in space, the geographic location and altitude of the receiver on Earth can be calculated to a precision of about 10 feet (3 m).

The applications of GPS are only limited by the imagination—it was used to track progress in the excavation of the Channel Tunnel between France and England, is now an integral part of the control of taxi and transportation fleets, and, as the FCC recognized, will be invaluable in locating people who are lost or in trouble.

developing nations to leap-frog wired communications and land directly in the 21st century. A wireless network has two very distinct advantages—time and cost. The infrastructure required to establish wireless communication is nowhere near as intrusive or expensive as stringing wires across the landscape. Expansion is merely a matter of adding to the capacity of existing base stations or putting in more of them, and the handsets are becoming ever cheaper as they are mass produced. Second, the time required for wireless installations is measured in months rather than years. Already in Latin America, for instance, it is predicted there will be as many wireless connections as landlines by the year 2006, and that the wireless network will put the Internet at the fingertips of more than half the population by 2010.

Virtual Reality

Virtual reality has captured the public interest for decades.

But it is no longer the stuff of science fiction. Soon, it may

transform computers into extensions of ourselves.

Ever since it was first mentioned in the science fiction novels of the 1980s, the idea of being immersed in a computer-generated world—or virtual reality—has captured popular imagination. Initial attempts at creating virtual reality, or VR, in the 1990s showed that the task of creating a convincing, useful visual world inside a computer was going to be difficult. But these efforts also spawned a vast array of types of computer-based imagery that will become widely used during the 21st century.

As a result, a range of emerging computer-imaging technologies is about to change the way we learn, work, and play.

AUGMENTED REALITY

One new imaging technology is augmented reality, where computer-generated images are superimposed on your view of the physical world, usually via see-through goggles. The

A VR HEADSET *like the one being tested below often incorporates stereo headphones in addition to visual information, immersing the user in a world of sound as well as 3-D vision.*

combination brings the power of the computer to a user's working environment.

For example, neurosurgeons would be able to see through brain tissue to the precise location of a tumor as they performed a procedure; engineers could view the stresses placed on a part in a machine as it operates; and golfers would be able to "see" the wind, or a precise contour map of a putting green, before taking that all-important shot.

As today's computing power and modeling techniques improve, the value of being able to create a space that you can enter and interact with in any way you choose will become more compelling for a host of users, including educational institutions and businesses.

The goal of any VR system is to create an illusion that sufficiently convinces your senses to make you believe you are actually inside the virtual space generated by a computer. The system must trick your senses by appealing to several of them at once, creating stereoscopic pictures that are slightly offset. One of the main problems facing VR is how to make the visuals more convincing and easier to use. The

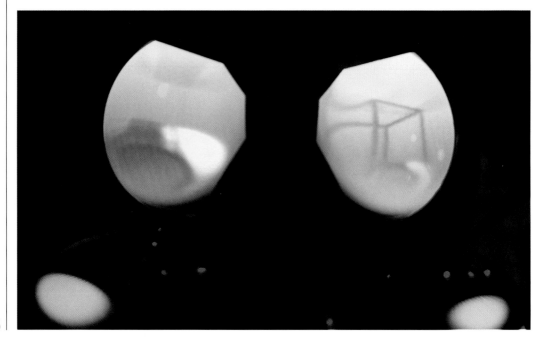

VIRTUAL REALITY *headsets enable users like geologist Peter Wilkness (above) to enter simulated environments, giving him the ability to operate in distant locations like Antarctica.*

low resolution of the current displays makes many users of VR disoriented or ill, and the headsets are bulky and cumbersome. But there are already ways of allowing a user to "touch" objects that are in a VR space.

Virtual reality holds special appeal to designers and engineers, who can create and test new objects in a VR space without the expense of creating a physical copy. For example, while it won't be quite the same as being there, plans are already under way to create virtual copies of landing sites on Mars, based on data captured by automated probes to the planet.

The applications of VR systems could be many and diverse. In courtrooms of the future, VR systems may take juries through a reconstruction of an accident or crime.

It may also be possible to meet and work in the same room as colleagues who are thousands of miles away. At least, that is the goal of tele-immersion, which uses VR techniques to allow people in distant locations to interact as if they were sharing the same space. Conference calls, job interviews and university lectures will never be quite the same again.

REACH OUT AND (VIRTUALLY) TOUCH SOMEONE

From the moment we are born, we use our sense of touch to explore the world around us, so it makes sense that if you want to feel part of a computer-generated world, you'd want to be able to touch it. But how?

The answer is "haptic interfaces," which use a physical force to give feedback to the user. The first forms of haptic feedback mechanisms were motor-driven arms that you inserted a finger into and which allowed you to run a fingertip over different objects in a VR space and feel the surface textures—smooth walls, sharp corners, rough spheres—all of which helps to create a convincing illusion that the virtual object is real. By reflecting the force of an object in a virtual space, haptic interfaces can be used to create an accurate simulation of the physical world, offering realistic training for dangerous situations. Some surgeons are already practising their techniques on virtual patients.

Researchers are working to give haptic interfaces different forms of sensitivity, such as temperature. They are also trying to find ways of increasing the scale of contact, from fingertips to whole hands and more.

Artificial Intelligence

It's entirely possible that the next 20 years will see the

development of a digital copy of the human brain.

Artificial intelligence, or AI, is a very broad field of computer science that is concerned with getting computers to do tasks that otherwise require human intelligence. Among other things, AI is the basis for robotics, computer-assisted learning, and data mining—the search for useful patterns of information in large amounts of data. Developments in AI are making computers easier to use by changing the way we interact with them. You can already buy systems that allow your computer to recognize spoken words and commands. More complex systems that are still at the design stage employ cameras and advanced imaging software that recognize and respond to gestures and facial expressions. Eventually, they may even learn to recognize your emotions.

AI is also helping us to learn more about ourselves. In trying to build artificially intelligent systems, we've realized there's a lot about intelligence that we don't yet understand. Some research groups are developing human-based robots in order to understand more about how our own brains work.

It remains a controversial issue as to whether or not a computer can be said to think at all, but AI systems have been developed that can beat the world's best human chess player, and control huge, complex manufacturing plants and refineries. Still more systems can design advertising posters, write music and poetry, and help doctors diagnose medical conditions. AI traffic systems can recognize when traffic jams are forming and change the timing of appropriate traffic lights to clear roads of congestion.

As computing power continues to increase and programming techniques improve, the sophistication and capabilities of the systems that are being developed are also greatly increasing. The old brute force computing approach to solving problems—using sheer processing power to crunch billions of numbers—has been replaced with some more sophisticated approaches that use mathematical rules of thumb to help a system solve a problem. And AI is being used to make products that are aware of their surroundings and can provide useful feedback to their users.

One form of AI is "expert systems," compressing the knowledge of a human expert in a particular field into an automated software system—the experience a lawyer uses to assess the likely outcome of a case, for example, or a financial analyst's ability to spot trends in the markets. Such knowledge-based applications of AI have greatly enhanced productivity in business, science, engineering, and medicine. Expert systems that can operate in real time are being increasingly used in many engineering applications, such as on board aircraft to diagnose and fix problems with instruments as they occur during flight, and in the metallurgical industry, where expert systems have been developed to help maximize the amount and quality of metal produced in blast furnaces.

But making an expert system is a slow and expensive affair, because you need to have the experts on hand to give you the knowledge

CREATING INTELLIGENT *robots is still a long way off. Jack the robot (left) is part of a project to develop a machine that learns by analyzing and exploiting its constraints. At the moment it is still learning to recognize the touch of a hand. It is hoped that MIT's Cog (right), one of the most sophisticated robots built to date, may one day possess the intelligence of a six-month-old baby.*

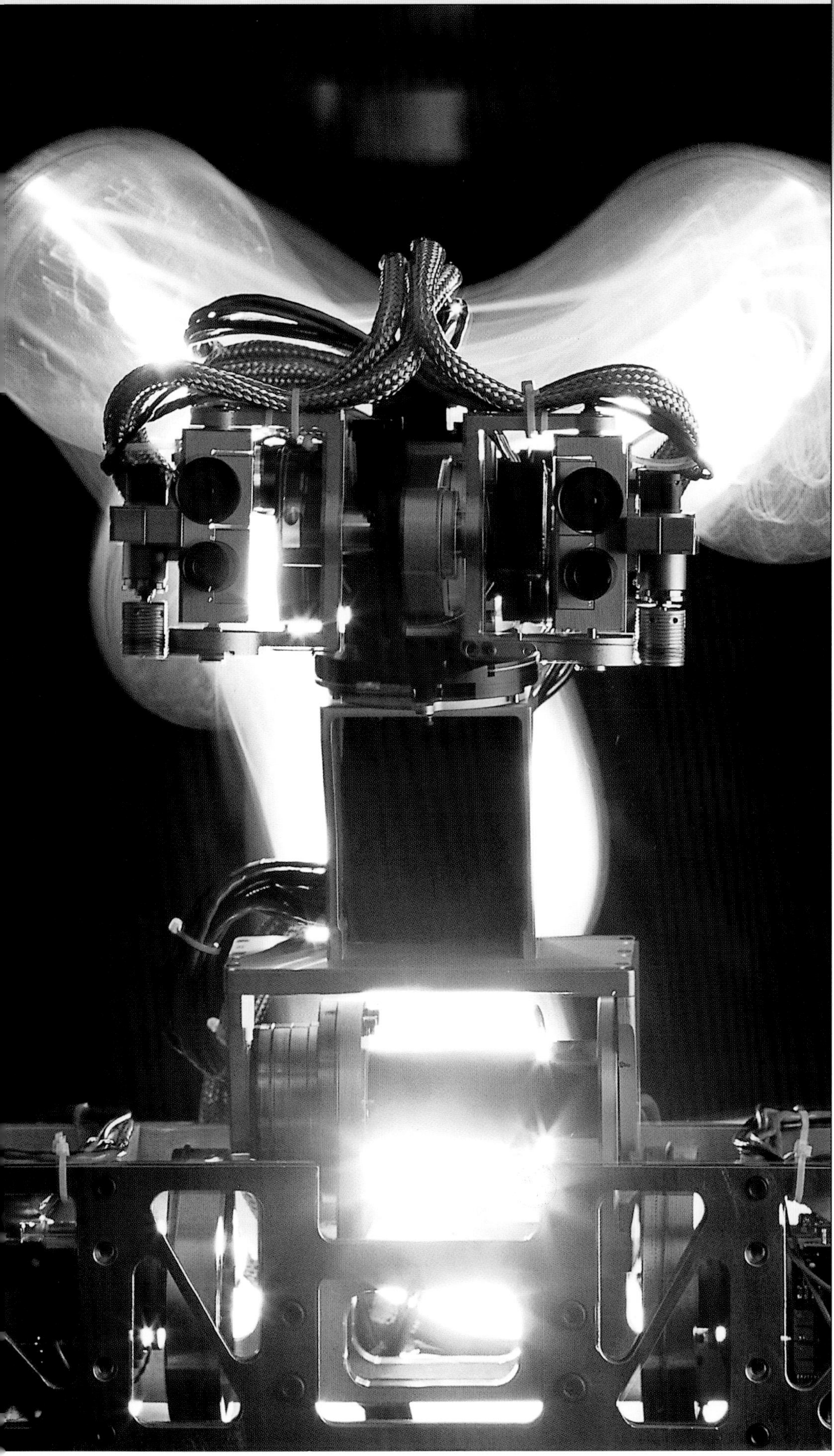

THE CUTE *R100 (right) is able to recognize faces, identify a few hundred words of Japanese, and obey simple commands. Its most important job, says its designer, is to help families keep in touch. It has an e-mail facility that delivers messages verbally.*

base in the first place, and creating efficient systems has proven to be a very difficult job that requires a unique approach for each application. What's more, because the information in most expert systems tends to be restricted to a very specific topic, the majority of them are highly limited in their use. For expert systems to perform competently over a broad range of tasks, they need more knowledge. Hence, the next generation of expert systems will require massive knowledge bases.

AGENT TO AGENT

The end of the 20th century saw an incredible growth in information technologies. Our lives are now so swamped with information that we require ever smarter tools to help us quickly find the information that is relevant to us and filter out that which isn't.

An agent is someone with expertise whom you trust to go out and act on your behalf. In AI terms, an agent is a software program, a personal assistant that learns about your needs and preferences and can reason, plan, and act autonomously on your behalf and within a defined scope. Around the world, developers are trying to create new agents of increasing ability and sophistication to find and interpret information. You may end up using dozens of personalized agents at work, all acting in tandem on a problem or a whole set of them, and only disturbing you when they require your approval for a decision they recommend. If you want to organize a business trip, you will use an AI travel agent to negotiate deals for plane tickets, rental cars, and hotel rooms, once you've told it when and where you're going. Agents are also starting to be used in business management and manufacturing. Journalists and financial services companies, for example, are keeping track of breaking news and the latest developments in their fields using agents designed to filter through online news sources. The computer company Sun Microsystems is using agents to plan shipping schedules. And as the complexity of industrial plants increases, and the tasks of human operators become more difficult, agents can be used to monitor production and warn operators of any problems.

A BETTER BRAIN

Most systems that are now called AI do not actually think for themselves or learn, they merely apply the rules that were programmed into them. There are many researchers who feel that the ultimate goal of AI is to make a truly conscious artificial intelligence. And many of them believe it's going to happen soon.

Computing power is set to continue to increase rapidly for the foreseeable future. Some researchers think that it's only a matter of time before machines are built that have far more computing power than the human brain. But how can you make a machine intelligent? One approach is to try and build

INTELLIGENCE IS *deceptive. SARCOS (left) may seem aware, but it does not sense the environment or react to it. It either follows a program or is controlled by an operator wearing a sensor suit.*

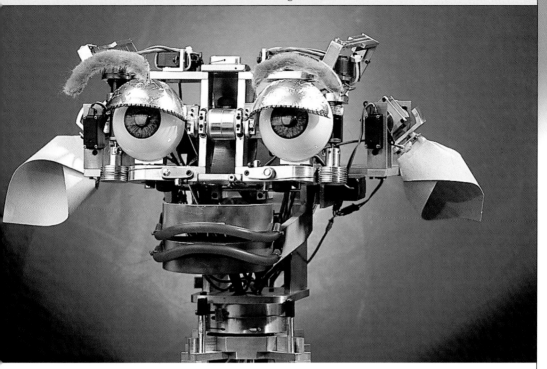

KISMET'S EXPRESSIONS *appear startlingly human (above), but its ability to react lies in the programming. The cameras behind Kismet's big blue eyes send data to its computer. The software allows the robot to detect people and bright toys visually or vocally. This then makes Kismet react by twisting its features into a look, such as the look of sadness seen here.*

an artificial brain, modeled on the real thing. Though much of the workings of the brain remain a mystery, we do know that, while a single neuron (or brain cell) is not intelligent, it is the immensely dense network of connections between billions of neurons that gives us our intelligent characteristics.

Seeking to mimic this effect, researchers have developed arrays of artificial neurons— devices known as "neural networks"—but with limited success. Two key factors have restricted the development of full-scale neural networks. The first is that the design of standard comput- ers is sufficiently unlike a brain or a neuron that researchers have had to develop new computer architectures in order to create adequate con- nections between each artificial neuron. The second restriction is the cost of building some- thing big enough to have the brainpower of a kitten, let alone a human being.

While some groups are trying to build an artificial human-like intelligence by copying the human brain, others are looking at the processes that led the human brain to evolve and are applying those processes to computer models to try to create an intelligence spon- taneously. These evolutionary algorithms use Darwinian rules of selection, recombination,

BRAIN COPYING: SHORTCUT TO IMMORTALITY?

Some scientists, like American Ray Kurzweil, believe it's only a matter of time before we are able to "reverse-engineer" human brains. Kurzweil's view is that, as medical imaging and sensory technology improves, it should become possible to accurately map the structure of a human brain. As we understand more about how the billions of neurons in our brains work and interact, it should be possible to create an exact physical copy of a brain, and for that copy to function just like the physical original.

This would mean that people could make "brain backups" of themselves at different times in their lives, just as writers make copies of docu- ments as they write them. If such a technology does emerge, our conceptions of consciousness and individuality might have to change radically.

mutation, and survival of the fittest to develop a system of data structures.

The advantage in this evolutionary approach is that the speed of computers allows processes that have taken thousands of years to occur nat- urally to be applied very quickly. Many genera- tions of programs can be run in a few hours. If, or perhaps when, artificial intelligences emerge that are smarter than us, should we feel threat- ened? Science fiction aside, there's no reason that something smarter than us would treat us as something to be disposed of. And it may be a remarkable opportunity for our development.

The E-Book

A world of literature and learning in a portable device

the size of a novel is what the e-book promises

to the readers of tomorrow.

Johannes Gutenberg invented moveable type over five hundred years ago, starting a revolution whereby the printed word replaced the oral transmission of information and cultural learning. Without a doubt, the invention of printing from moveable type was a transforming moment in modern history.

Today, according to information technology leaders, the world is once again on the brink of a momentous revolution in the transmission of printed texts, a revolution that has already been termed "the dawn of the age of e-books."

THE BEGINNING

Already a reality are electronic books that are the approximate size and weight of a meaty novel. They are wholly portable but can contain a whole cache of novels, reference books, and dictionaries. They even carry the correct pronunciation of challenging words or names within a text. These e-books allow the reader to change the size of the type and the orientation of the page at the press of a button. They can also be read in the dark. In fact, they are common enough to have caused the digitization of thousands of titles, including many of the classics and the latest releases of some of the world's leading fiction authors.

But even so, the electronic books currently available on the market are believed to be simply the crude prototypes of the text revolution to come. According to proponents who are backing the technology, e-books will constitute half of all "book" sales within the first few decades of the 21st century.

While the hardware has been around for some time, the weaning of book buyers into this new medium has been relatively slow. It has been held up by several factors. The first is the widespread skepticism that readers would will-

ingly give up the familiarity and sensuality of the printed page. Then there is the sheer labor involved in reading large passages of text from monitors or "grainy" liquid crystal screens. Moreover, the human brain and eye need the encouragement of certain time-honed formatting conventions to achieve what is known as "seamless word recognition."

Sensing the inevitable direction of text in the future, however, some of the world's leading developers of information technology have been trying to iron out all the kinks in type fonts, text formatting, and screen resolution—"getting the reading software to catch up with the writing software." As these developments progress, the promise is of an e-book medium that appears "sharper," and with pages that are "more print-like."

ON THE DRAWING BOARD

At the start of the 20th century, British author H. G. Wells envisaged the possibility of an electronic book. But the e-books that are on the drawing board for the future would amaze even that far-sighted science fiction visionary.

Imagine a single electronic book that can download copious volumes from the Internet in no time; e-books with the capacity to store entire libraries; and e-books that allow readers to instantly cross-reference any queries that arise about words or facts via in-built encyclopedias, dictionaries, and numerous hyperlinks. The technology now on the drawing board promises to revolutionize educational processes and, almost inevitably, the function of the traditional institutional or public library.

With the continuous refinement of audio technologies in

JOHANNES GUTENBERG'S *most significant innovation in the world of print was the efficient molding and casting of moveable metal type (left) at a time when secular scribes simply could not keep up with the ever-increasing demand for books.*

E-BOOKS *like the REB1100 and REB1200 (above) that promise to "shed a whole new light on reading" can carry up to dozens of novels, newspapers, and magazines at one time. Some e-books will have the capacity to store entire libraries, raising questions about the future of libraries like the one at Trinity College in Dublin, Ireland (top).*

giants, Random House, recently announcing that any authors of e-books published by them will receive 50 percent of sales revenues compared with the 10–15 percent that they generally receive for print sales.

For the publishers of these new media, there are further issues to be resolved concerning the provision of failsafe encryption technology to protect the privacy of the end user, who may never again find it necessary to visit a bookstore to browse the latest releases. Sample text and single chapters of new titles are already available to view on the Internet.

these e-books, there is also the suggestion that for those who are visually impaired, and for those who simply find reading text visually a chore, the e-books of the future will have the facility to read themselves aloud.

Other hurdles to be surmounted on the way to the popular uptake of electronic books concern the question of copyright for authors or "content providers," to ensure that they receive their deserved income. It is highly likely that e-book authors will receive much higher royalty payments, with one of the world's publishing

The main question, of course, is whether readers will ever let go of the pleasure of leafing through lush, coffee table volumes with glossy illustrations, or give up the sheer convenience of reading softcover publications on the bus, in the bath, or on the beach.

If we judge by the growth of Barnes and Noble—one of the world's leading online book distributors and already the fourth largest e-commerce site in existence—the future of reading is most decidedly electronic.

Cyber War

In the age of cyber war, there is no front line. And when

nations go to war with computers blazing, we are all targets.

Something mysterious happened one night in 1991. During the Gulf War, over 120 Iraqi fighter planes flew to Iran, where they sat out the conflict on the ground. Why?

No one knows for certain. Intelligence experts suspect that the planes decamped after computer hackers from an enemy nation, or nations, inserted a computer virus into the central command-and-control system of the Iraqi Air Force. The virus knocked out the system, leaving pilots effectively "blind" and commanders "silent." With all communications dead, it was best to avoid battle and fly to the safety of a neutral nation—or at least that's how the theory goes.

If true, this was the first known instance of cyber war: the use of the microchip and computer networks to attack enemy defense systems. It is also the shape of things to come.

Experts predict that future battlefields will be on land, on sea, in air, and in cyberspace.

A MILITARY REVOLUTION

Cyber war is now possible, due to a revolution in military technology. It is a revolution based upon computer power, not firepower. Experts state that the impact of computer technology on military operations has been far greater than that of past military innovations such as gunpowder, the English longbow, aircraft, tanks, missiles, and even nuclear weapons. Computer technologies have fundamentally altered the way in which advanced armies now organize themselves and conduct war. This is because

THE GULF WAR is widely believed to be the first war in which a cyber assault was successfully launched against an enemy nation, allegedly grounding over 120 Iraqi fighter planes.

the integrated application of these new technologies has led to unparalleled improvements in many areas—in weapons, in delivery systems, and in battlefield communications.

Enemy positions are now more easily and safely identified. Weapons are increasingly accurate and are launched at a good distance from their targets, and the number of troops who are exposed directly to enemy fire is greatly reduced. This mix of precision hits and low casualties is known as "distance warfare," and it is a key characteristic of cyber war.

The 1991 Gulf War is a case in point. At the time of the conflict, the anti-Iraq coalition possessed weapons packed with microchips: radar-guided missiles, radar-killing missiles, and laser-guided bombs. They launched these and other smart weapons from a distance or released them from crewed and uncrewed planes. And the planes were loaded with computer-enhanced operation, communication, and navigation capability. Back home, powerful computer networks enabled the high command to oversee the conflict from Washington and London. This was possible because tactical and strategic communications equipment had been sent to the Gulf. Ultimately, hundreds of thousands of personnel were linked through the system. Orders were given instantly, based upon "real time" information from the war zone.

Since the Gulf War, smart weapons, communication networks, and distance warfare tactics have been applied to conflicts ranging from the civil war in the former Yugoslavia to the war against terrorism in Afghanistan.

DISTANCE WARFARE *was a very real feature of the United States-led Operation Desert Fox against Iraq. United States Navy fighter planes like these readied on the flight deck of the USS Enterprise were armed with increased computer capability and smart weapons that could be launched far from the target.*

THE POWER OF ONE

Stories of teenage hackers breaking into sensitive computer systems or vandalizing websites are stories familiar to all of us. Such attacks are real, but they are not part of cyber war. They are not coordinated attacks against an enemy in nation-to-nation conflict. They do, however, illustrate the power that an individual can command at the click of a mouse.

In cyber war, a few clever people, doing clever things, can wreak havoc. Worse, their handiwork may not be obvious until it is too late. Communications are disrupted, weapons fail, or military intelligence is altered or stolen from a "secure" computer network, with devastating consequences. This is known as "asymmetric warfare" because a handful of cyber soldiers, with a technological edge, can outmatch a much larger adversary.

The arsenal of the new warrior is stocked with a wide range of weapons, such as logic bombs, computer viruses, packet sniffers, and system overload software. Spoofing tools enable cyber soldiers to insert malicious codes or false information into a system. That includes embedding morphed video images into foreign television broadcasts. Skilled cyber soldiers can download their weapons from the World Wide Web with ease. Then, thanks to the Internet,

CYBER HEADQUARTERS, *far from the battlefield, keep commanders in touch with their troops in "real time" (above), a far cry from the old control rooms that aimed to keep guns and troops moving toward the front (left).*

they can launch anonymous attacks from anywhere on the globe. Policy makers and politicians have been slow to recognize the danger of enemy hackers. They tend to attribute hacking to bored teenagers, or corporate crooks. To highlight the real risk, the United States National Security Agency (NSA) staged an exercise called "Eligible Receiver." Over a two-week period in 1997, a crack team of almost 70 NSA hackers downloaded freely available intrusion software from the Internet and used it to break into allegedly secure computer systems. One system controls the nation's electrical power grid. Another was the command-and-control system of the military's Pacific Command. The hackers could have unplugged the entire country and crippled the Pacific Command and its 100,000 troops. Worse, they were elusive. Government specialists tracked down just one small group of intruders.

WE ARE THE TARGET

Cyber war promises generals undreamed-of firepower while appealing to leaders who are

averse to soldiers coming home in body bags. But cyber war is a double-edged sword. Our enemies may deploy the same weapons that we use. And that makes each of us a target simply because advanced military forces use the same telecommunications network as ordinary citizens. So an attack against enemy military networks is, in effect, an attack on civilians.

An American think tank, the RAND Corporation, estimates that 95 percent of all military communications in the United States travel over civilian telephone lines, satellites, fiber-optic cables, and microwave links. In fact, RAND claims that nearly everything the military does depends upon civilian systems. That includes paying, training, equipping, and mobilizing soldiers, aiming missiles, and designing weapons. Other highly computerized countries are vulnerable too, and for the same reason.

We seldom notice them, but networked computers run the banking, air traffic control, and emergency response systems of most countries. They keep the power lines humming, the telecommunication links open, and water and sewage moving. Morever, many of these systems are interconnected. Together, the telephone cables, optical fibers, communication satellites, and networks of powerful computers are known as a nation's critical infrastructure.

Yet despite the vital importance of this infrastructure to military, government, and private activities, operators often skimp on security or backup systems to reduce costs. This ultimately means that cyber targets are everywhere. And the attack would hit home, quite literally. Imagine if automatic teller machines randomly debited and credited your account, or traffic lights changed haphazardly. What would happen if your name popped up on a list of convicted felons, or your local television lost control of its programming and aired misinformation? Imagine if all the faucets went dry, the sewers backed up, trains, planes, elevators, and automobiles were immobilized, and the stock market crashed. All it takes is some skilled hacking.

THE NEW HOME FRONT

In January 1999, the then U.S. President Bill Clinton announced a multibillion-dollar program to protect the nation from the threat of cyber war. The main initiatives included more research into ways of detecting attacks, warning government agencies of an attack, and encouraging the private sector to do the same.

These projects were on top of the existing cyber defense structure, headed by the Critical Infrastructure Coordination Group (CICG). Under the CICG umbrella come various operational and policy centers, all working to safeguard the country's critical infrastructure.

The task is enormous. And it begins with nearly every computer in every critical network, both military and civilian. Few are truly secure. Most operate on commercial off-the-shelf operating systems. Surprisingly, buyers seldom demand full documentation of the codes inside those systems. Do they contain hidden instructions? Are there "back doors" through which hackers can sneak? Instead of getting answers, however, network operators usually turn to "intrusion detection," the ability to find out if a computer has been attacked. Many experts claim this is too little, too late. They say it's essential to build secure computers that keep out hackers, but that's expensive.

THE MILITARY USES *the same information networks as civilians an estimated 95 percent of the time. A cyber attack on the military, therefore, is an attack on civilians and could bring down everyday systems such as traffic lights, creating chaos.*

Then there are human obstacles. Because their networks are interconnected, military, government, and corporate leaders must cooperate for effective defense. Yet they have proven reluctant to reveal their weaknesses or secrets. Privacy groups fear that if leaders do cooperate, that will pose a threat to personal privacy: No one wants Microsoft or the FBI monitoring his or her daily activities.

Finally, despite efforts by the CICG, the question of who is responsible for repelling cyber invaders—army, navy, air force, FBI?—is left unanswered. The United States is not alone in its struggle to resolve these issues. Any nation dependent upon computers is a target. Cyber war is not science fiction. It is science fact. The question is, are we prepared?

ACADEMICS AND LEADERS *of the computer technology industry met with then United States President Bill Clinton in 1999 to propose a new Internet security plan as a result of hacker attacks on some of the leading commerce sites.*

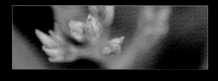

CHAPTER SIX

THE LIVING WORLD

In nature there are no rewards or punishments;
there are consequences.

The Face of Clay,
HORACE ANNESLEY VACHELL (1861–1955), British writer

The Bountiful Seas

The seas make up 95 percent of the planet's biosphere,

and we are responsible for fishing them to extinction.

Every day, at least 38,000 factory ships are roaming our waters in search of an ever-dwindling catch of wild fish. The seas have provided us with food for thousands of years, and just a few decades ago it seemed this resource was unlimited. Yet today, 13 of the world's 17 major fisheries are in steep decline and several others have collapsed, including the cod and haddock populations off the east coast of the United States and Canada. There are other indications of diminishing fish stocks. For instance, the total number of fish caught per year, which reached its global peak in 1989, has not increased since then, despite some major advances in the technology for finding and for catching fish.

Apart from continuous overfishing, the world's fish stocks face other, nonhuman pressures. El Niño, the southern oscillation that warms the sea, may be a factor in reducing the number of juvenile fish that make it through to adulthood. An unusual warming of waters off the coast of Alaska in 1976 is believed to have triggered important biological changes that are still in effect, increasing the Alaskan salmon catch yet reducing the salmon catches from California waters. Global warming may in the future exacerbate these uncertainties.

For people the world over there is cause for concern. Those living in developing countries are particularly reliant on a diet high in fish. And in industrialized countries like the United States, as people look to a healthier diet and as land for cattle ranching becomes increasingly scarce, more and more people are turning to seafood. Fish and beef have long been the two major sources of protein, but since 1990 there has been very little growth in the production of either. Yet by 2020, it is estimated that the Earth's human population will have grown by another 1.5 billion. A collapse of world fisheries could spell disaster for hundreds of millions.

ADVENTURES IN AQUACULTURE

As fish stocks decline, there will be a growing demand for fish bred in captivity. Aquaculture is expanding faster than any other sector of the world food economy, with production almost tripling between 1990 and 1998. Today, one in five fish consumed comes from a fish farm, but that figure will be one in two by the year 2025.

However, the aquaculture industry will have to plan carefully if it is to avoid creating more problems than it solves. With such a demand, salmon are now being bred to grow faster and mature earlier. The first genetically modified

WILD FISH *stocks are dwindling and catches like this one (right) will become increasingly hard to come by. To meet the soaring demand for seafood, salmon (above) are being bred in captivity to grow ten times faster than their wild counterparts.*

FACTORY TRAWLERS *like the one shown above have driven many popular species of fish to commercial and even biological extinction. Now, aquaculture farms like this salmon farm in Scotland (right) hold promise as future sources of protein.*

Atlantic salmon grow ten times faster than their wild counterparts, but they have not yet been approved for human consumption. While these fish are good at growing, they aren't necessarily good at long-term survival. There is evidence that some captive-bred fish have escaped and interbred with wild stocks, passing on their genes and potentially weakening the gene pool of wild salmon. Farmed salmon are vulnerable to diseases such as sea lice and infectious salmon anemia, which can affect wild populations.

There are environmental problems with farming salmon and shrimp (prawns), too. According to one report, the waste produced by farmed salmon in Norway is roughly equivalent to the sewage produced by Norway's four million people. Also of concern is the clearing of mangroves to make way for shrimp farms, since mangroves protect coastlines and provide nurseries for juvenile wild fish.

A further problem is the diet of wild-caught small fish that are fed to captive salmon. It takes up to five tons (tonnes) of these wild fish—such as herring and anchovy—to produce one ton (tonne) of salmon, putting considerable pressure on fish stocks further down the food chain.

Herbivorous fish farming offers a far more efficient way of converting raw materials into protein. Carp and catfish, for example, require around 3.5 pounds (1.6 kg) of grain for every 2.2 pounds (1 kg) of weight gained. In China, carp have been grown in combination with rice. As well as being an efficient use of land, the growing rice helps to break down the fish waste. Mississippi produces about 60 percent of the catfish in the United States, grown in 174 square miles (450 sq km) of catfish pond.

MANGROVES (above) provide nurseries for juvenile wild fish and are a natural protection for coastlines. Many of the coastal ponds where shrimp are farmed are created by clearing such mangroves, causing significant environmental problems.

WILL FISH STOCKS EVER RECOVER?

There are two main approaches to protecting and restoring dwindling fish populations. One strategy is to attempt control over the size of the world's fishing fleet. The other is to create marine reserves where fishing is prohibited, in order to give populations time to recover.

There is mounting evidence that marine reserves, like the one created on the Georges Bank off Massachusetts in 1994, can be highly effective in restoring depleted ecosystems. The scallop population there is already recovering, the fish are larger than they were and spawn more frequently. Great Britain has committed to creating more marine reserves in the near future, which will be supported by the industrial sector. Such reserves will help to protect fisheries while the slow process of thrashing out sustainable fishing policies goes ahead.

Many experts believe that controlling fleet size is the only way to avoid fishing the oceans to extinction. Members of the United Nations Food and Agriculture Organization plan to

BANKING ON THE FUTURE

A plan is in place to protect the fish that provides us with of one of the greatest but most endangered luxuries: caviar. Caviar is the eggs of the sturgeon, a large fish that lives in the Caspian Sea. Polluted by local factories, the Caspian is dying. Biotechnologists from the Food and Agriculture Organization have set up a gene bank to preserve the sturgeon's tissue and sperm on ice. Similarly, gene banks have been set up across Europe, Canada, and the United States to store the frozen sperm and eggs of other commercial fish species that may be at risk, in the hope that the fisheries can one day be reestablished.

adopt measures to control the size of the long-distance fishing fleets. Japan began this process in 1999 by scrapping 130 vessels.

DRUGS FROM THE SEA

There are farther-reaching effects of a dwindling fish supply. On the rocky headlands in southeast Australia, marine biologist Kirsten Benkendorff has discovered one of the sea's precious secrets. Her work has focused on one particular sea snail. Over a period of four years she extracted hundreds of compounds from this small snail—a painstaking process—and analyzed them for their antibiotic properties. In so doing, she isolated a compound from the sea snail's eggs that has quite different chemistry from that of any known antibiotic. The discovery could provide a much-needed answer to the problem of bacterial resistance to available antibiotics.

Until recent times, all our medicines came from the land and the forests. Soil-dwelling organisms have given us antibiotics such as penicillin and amoxicillin, which have saved millions of lives. But medicines extracted from sea life are far rarer. Research is now revealing that some of the ocean's strangest and most obscure creatures contain substances that could one day be the basis of cures for cancer, AIDS, arthritis, and many other chronic illnesses. For example, the blood of the horseshoe crab—which crawls up onto United States beaches from Maine to Florida each spring to lay its eggs—has given us a simple test for a dangerous toxin produced by salmonella bacteria.

The United States leads the way in research now going on to isolate compounds from sea creatures that could be medically important.

NO ONE KNOWS *how many species of sea creatures holding medicinal benefits, such as the horseshoe crab (below), may have already been lost to commercial fishing.*

SCRAPING THE BOTTOM

In their efforts to find fish, commercial fishermen are putting out longer, deeper nets like the ones illustrated above, fishing farther down the food chain and moving to untapped areas such as the Southern Ocean, south of Australia and New Zealand. A recent target is the Patagonian toothfish, which lives up to 1 mile (1.6 km) below the surface and can grow up to 7 feet (2 m) long. In the 1990s it was barely fished because it lives at such a great depth and in areas remote from the major fishing ports. But while little is known of its biology, in the early years of the new millennium the species began to be fished beyond its limits. Almost four times the current quota are caught. Many are poached by Spain, Norway, and Korea. The toothfish could be commercially extinct within a few years.

Researchers at the University of Hawaii have discovered that cryptophycin, a chemical found in blue-green algae, alters the internal structure of cancer cells and prevents them from spreading. Other research has found that a neurotoxin from a Pacific sea snail is a potent painkiller and could become a viable alternative to morphine.

There are many more sea creatures with potential pharmaceutical value still being discovered today. Researchers in Australia have recently found hundreds of new species of sea fans, corals, and sponges living on underwater volcanoes (known as seamounts) between New Caledonia and Tasmania.

From a sample of just 25 seamounts they found a wide range of more than 800 species, with each seamount community quite different from the next. Unfortunately, because of their high biodiversity, seamounts are increasingly becoming a target of deep water trawlers, and researchers are calling for an international approach to protect them.

183

Global Warming

Unraveling the complex interactions between Earth's oceans,

atmosphere, and land masses is a challenge for science, but

necessary if questions about global warming are to be answered.

Looking back over the past 1,000 years of global air temperatures, scientists have noted a remarkable warming trend in the last 50 years of the 20th century. After natural factors are taken into account, it seems clear human activity such as deforestation, modern intensive farming practices, and land degradation have been contributors. But by far the biggest culprit has been the burning of fossil fuels by an increasingly industrialized world.

Since the Industrial Revolution in the 18th century, human society has become ever more urbanized and mechanized. Technology has allowed humans unparalleled control over the environment, but has also caused large amounts of industrial gases to pour into the atmosphere: carbon dioxide, chlorofluorocarbons, nitrous oxide, and methane. These gases, now collectively known as greenhouse gases, allow much of the Sun's shorter-wavelength radiation into the atmosphere. But not all of this heat, in the form of longer-wavelength radiation, is re-radiated back out into space. The gases trap some of the heat and make the atmosphere behave rather like a greenhouse. As greenhouse gases increase, air temperatures rise. Leading environmental scientists have predicted that, as a result, global temperatures by the mid-21st century will rise between 2°F and 4°F (1°C and 2°C). This is on top of the 1.3°F (0.7°C) increase recorded during the 20th century.

FUTURE EFFECTS

Our oceans can store vast amounts of this heat energy from the Sun, slowing down the atmospheric warming process. But gradually, this stored energy feeds back into the world's weather patterns, possibly in the form of more frequent El Niño events (persistent oceanic pools of warm water that are bigger than the largest continents). This built-up energy is also believed to interfere with the flow of massive deep ocean "conveyor belts"—slow-moving circulations of water that meander around the great oceans of the world and help redistribute heat. Sudden changes in the conveyor belts 11,000 years ago are thought to have triggered quite dramatic climate changes that led to a number of species becoming extinct.

The Atlantic Ocean's conveyor system, believed to create the relatively warm European

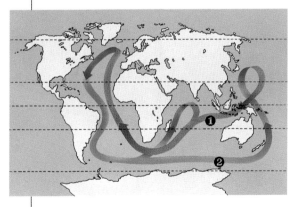

THE GREAT OCEAN CONVEYOR BELT *is a system of deep, cold currents and shallow, warm currents that circulates the globe. The ocean current that brings warm surface water north and east and heats Europe is called the Gulf Stream. With global warming, melting ice from the Arctic and Greenland could push fresh water into the North Atlantic, disrupting the flow and density of the current and changing the path of the Gulf Stream. If that happens, Europe will get very cold, and other ocean currents connected to the Gulf Stream will be disrupted. The map above shows: 1. Shallow, warm currents. 2. Deep, cold currents.*

EL NINO AND LA NINA

An El Niño occurs when a large area of warm water accumulates in the equatorial Pacific Ocean off Peru. This persistent warm water anomaly alters large-scale atmospheric circulations around the world, bringing severe drought to some areas, such as eastern Australia, and floods to other areas, such as the western coastline of South America. When sea surface temperature anomalies are reversed, a La Niña is said to occur, and in most cases areas of flood and drought are reversed.

THE BURNING OF FOSSIL FUELS *releases large amounts of carbon dioxide into the atmosphere which creates a kind of cushion around the Earth. Long-wave radiation hits the cushion and can't escape, which is why the Earth gets hotter and hotter.*

climate, transports water equivalent to 100 Amazon Rivers. Sudden massive injections of fresh water from melting icebergs or increased rainfall are thought to have altered this circulation in the past, and global warming could trigger such an episode in the future. The flow-on effect would be a sudden drop in North Atlantic winter temperatures of around 9°F (5°C).

With the expansion of warmer oceans, coupled with increased water runoff from snowlines and retreating glaciers, comes the threat of rising sea levels. Global sea levels have already crept up by about 4 inches (10 cm) during the last 50 years of the 20th century, with accurate satellite altimetry technology recording a rise of 0.12 inch (0.3 cm) per year in just the last eight years of the century.

Toward the end of the 21st century, rising sea levels are likely to present a major challenge to people on low-lying land, particularly on river deltas and Pacific Ocean islands such as those in Kiribati and Tuvalu, which are barely 6 feet (1.8 m) above sea level. The predicted rise is almost 8 inches (20 cm) by 2050, which is faster than coral reef can grow; this will leave low-lying land increasingly exposed to erosion,

raging seas, storm surges, and the increased climate variability that global warming brings.

The result may be storms that are more frequent and intense, triggered by the extra heating, and with more variable storm paths that bring more violent weather systems to areas that were rarely affected in the 19th century. It is expected that more, and possibly stronger, El Niños and their alter egos, La Niñas, will occur, bringing with them the grim effects of famine, landslides, water shortages, and massive floods to increasing numbers of the world's burgeoning population.

People will by no means be the only victims of global warming. As regional climates change, many endangered species that are already battling for survival in remnant pockets of their formerly extensive habitats will come under increasing stress, too. Isolated as they are by expanding urbanization and industrialization, endangered species will be unable to migrate to new ecosystems. Even relatively short-lived climate variations are likely to push entire species to extinction. Plants are also greatly threatened by rapidly changing climates that could render their present location unsuitable for regeneration. And as human communities become affected by crop failures and famine, they will be pushed to clear residual habitats in an effort to survive—putting further pressure on an already stressed ecosystem.

LOW-LYING REGIONS *of the world like Florida in the United States will be devastated if global warming and sea level rise continue at their predicted rates, and the tropical islands in Sulawesi, Indonesia (above) could be completely submerged.*

GLACIERS

Glaciers take a few years to respond to any changes in their local climate and are therefore a good natural global thermometer, providing tangible evidence of global warming. The once extensive glaciers on Irian Jayan peaks on the Pacific island of New Guinea have almost totally vanished, while the Bering Glacier in Alaska has dwindled by 6 to 8 miles (10 to 13 km) during the 20th century. And the Larsen B Ice Shelf in Antarctica collapsed for the first time in recorded history in 1999, having been stable for at least the last 400 years.

TECHNOLOGY—TOMORROW'S HOPE

Technology is often painted as the villain in the global warming debate, but it may also provide a way to partially ameliorate the devastating effects. Advances will focus on trying to counter the causes of global warming, as well as those that aim to protect life from it.

Around the world, efforts are under way to reduce greenhouse emissions by focusing on fuel efficiency, alternative energy sources, the encouragement of mass transport over the use of the motor car, and carbon entrapment techniques. Urban areas will need to become less energy-intensive, powered by non-polluting, renewable energy sources, such as solar power and wind. Refinement and expansion of recycling and whole-of-life product planning will be needed to optimize use of waste that was burnt or buried throughout the 20th century.

Protection strategies are every bit as diverse. Wildlife corridors are being established, which link isolated wildlife reserves to facilitate migration. Possibly the most ambitious of these to date is the plan to link national parks in the Yukon and Yellowstone in the northwest of North America.

Biotechnology and genetic engineering techniques are developing a new range of plants and animals that are resistant to salinity, drought, and disease, which could replace traditional crops and herds in modified regional climates of tomorrow, to feed the rapidly expanding global population. Also being refined are hydroponic techniques which should

ensure a more economical use of scarce supplies of fresh water in regions where shortages are exacerbated by local changes in climate.

CHANGING LAND USE

Improved management of our vast land masses provides us with numerous opportunities, right down to a community level, to slow the progress of global warming, often with other ecologically beneficial spin-offs. Replanting and ongoing harvesting of new forests and restoring wetlands will help to capture greenhouse gases, since young trees are excellent at absorbing carbon dioxide in the air. Using plantation wood in place of fossil fuels like oil or wood from old-growth forests will also help reduce the emission of greenhouse gases. New, ecologically sound techniques for managing crop and grazing land are becoming available that reduce land degradation and minimize the amount of fertilizers needed to extract crops from the land.

THE FINANCIAL WORLD

Even the financial markets have a role to play. With carbon credits trading, both developed and developing nations will be given a limited quota of allowable carbon dioxide emissions to support their industrial activities. These quotas, based on population numbers, will be able to be traded and sold between countries, creating an economic value for clean air. An inefficient, heavily polluting factory will need to buy more carbon credits, while a modern, efficient, and less-polluting factory may avoid such costs,

CHESAPEAKE BAY

The impact of modest sea level rises is vividly demonstrated by the loss of coastal ecosystems in Chesapeake Bay in the United States. A combination of slowly subsiding land (unrelated to global warming) and greenhouse-linked rising sea levels have subjected this region to an effective sea level rise of 0.14 inches (0.35 cm) per year throughout the 20th century. This is similar to the rise predicted globally for the entire 21st century. Between 1938 and 1979, one-third of the Blackwater Wildlife Refuge (above) was lost to inundation. Studies by the United States Environmental Protection Agency show that a 20-inch (50-cm) sea level rise may lead to a loss of 17 to 43% of United States coastal wetlands.

giving it a price advantage for the products it produces. In this way, industries in the wealthier nations—which are responsible for most of the greenhouse gases—will be under pressure to reduce their emissions, or else have to bid with other companies for the limited number of "pollution quotas" in the market.

Although this may have some negative impact in some industries at the onset, international agencies such as the World Bank know it is much cheaper to prevent natural disasters triggered by global warming than it would be to respond to them. All of these programs are being driven by internationally agreed objectives contained in the United Nations Framework Convention on Climate Change.

Industrial Ecology

Will it be possible to create an industrial ecological vision for

the 21st century that actually works? One thing is for certain:

We will need to try.

The failings of industry to date have been created by the inadequacies of the economy in which it operates. It has been cheaper to pollute and waste energy and resources than to be responsible with resources and not pollute at all. With those resources now rapidly dwindling and with the effects of pollutants reaching crisis point, it is now imperative to turn industry around, and that cannot be done without restructuring the associated economy.

New Industry, New Economy

Creating a new economy where environmentally responsible behavior is rewarded through profit and production incentives has been championed by institutions such as the Rocky Mountains Institute in Colorado in the United States. They have devised Natural Capitalism, an economic model that has

MANUFACTURERS OF THE 21ST CENTURY *are beginning to take new approaches to operations. By paying attention to the control of waste and pollution in the design and then production of their products, they may attain optimum efficiency.*

already had some success at implementing new economic and industrial principles.

Natural Capitalism factors into a production process the value of every resource used and the cost of every service, including those that nature provides for us. We cannot live without ecosystem services such as control of the composition of the atmosphere, the cycling of nutrients, the cycling of water, and the stabilization of climate, yet these are rarely factored into any economy.

Bees are a simple example of our careless attitude towards natural services. Without bees, most crops would not be pollinated and would fail. There is no affordable substitute for bee pollination of crops, and it is an essential part of agriculture. However, few agricultural economies factor in the costs and environmental requirements for this pollination service.

Sustainable Development

Increasingly, we are stripping non-renewable resources from the Earth and will one day run out of these materials. This includes not only our use of minerals and metals but also our use

of so-called renewable resources such as clean water, native forests, and sea life. If we use those resources faster than they can be replaced, they too will eventually be depleted. A much more economical industrial system would be one that does not use resources faster than they can be renewed, that recycles as many non-renewable resources as possible, and that values all the elements in the production cycle.

The concept of sustainable development is based on the careful husbandry of precious resources. Sustainable development can be defined as development that meets the needs of the present without compromising the ability of future generations to meet their own needs. Many of the issues at the center of sustainable development, such as air quality, water quality, and biodiversity, are global in scope and require the attention of all nations.

Producing Waste

Waste is the product of the industrial process that has not previously been accounted for. It takes energy and materials to produce waste, and it costs money to dispose of it. Because waste is expensive to produce and deal with, it is more economic to design waste out of a production system. A truly efficient industrial technology produces no waste at all.

Currently, in most developed economies, the weight of material extracted from the ground, used in production processes and ultimately disposed of every day is around 20 times the body weight of each person in

NATURAL CAPITALISM *factors in the cost of any resource or service used in a production process, including those provided by nature. In this way, the value of the pollination service offered by a bee, for instance, can be properly assessed.*

that economy. The flow of resources worldwide is in the order of a half-trillion tons (5.5 billion tonnes) per year. Of that, only about 1 percent ends up in durable goods; the rest is wasted. Current production systems are approximately 99 percent waste.

An example of waste hidden in the system is the energy industry. The efficiency of converting coal at a power station into light in a room, for instance, is under 3 percent. Power plants in the United States throw away the same amount of energy in waste heat as Japan uses for all its energy needs. Cars use less than 1 percent of their energy to move the driver. In fact, most modern, Western economies are less than one-tenth as efficient as the laws of physics permit.

This extravagant waste of energy is simply unnecessary, even with current technologies. Relatively small, cost-effective changes to most existing buildings could improve their energy and water efficiency by a factor of at least three or four. New buildings can be designed for energy savings closer to a factor of ten. That's a 90 percent reduction in energy consumption and costs, with better performance and lower construction costs.

Toxic wastes are a particular problem that can be designed out of most production processes. In the majority of cases there are non-

toxic alternatives, and often there is no need to have toxic chemicals in the cycle at all. Increasingly, the old methods that relied on the cheap disposal of toxic wastes to be economic are being reformed. It is increasingly difficult and ever more expensive to dispose of toxic wastes, which is why smart operators are learning to manage without them.

MATERIAL AND ENERGY FLOWS

The emerging science of industrial ecology uses conceptual tools to analyze and optimize the flow of energy and materials within production systems. It follows the production cycle from extraction through production and use to disposal of materials, products and energy, and tries to improve these "flows." Often the improvements in these flows are simple: Considering how materials and energy can be redirected into different uses, for example, or how products can be designed for reuse before they are even built.

FEEDBACK

Dr. Amory Lovins of the Rocky Mountains Institute argues that feedback loops within production processes will produce environmentally sustainable industry. If the people who are responsible for creating waste and pollution are directly affected by the presence of those products in the environment, they will find better designs for their production lines that eliminate those wastes. For example, a car's exhaust would have to be much cleaner

HUNTER AND AMORY LOVINS *(above) are the research analysts and founders of the Rocky Mountains Institute. They have shown that advanced techniques for resource productivity can greatly, and competitively, reduce environmental impacts.*

if the tailpipe were directed into the passenger compartment rather than at pedestrians. Factories would be cleaner if their water intakes were downstream of their outfall, or if their air intakes were connected to their smokestacks. Lovins asks, if you wouldn't release your waste products onto yourselves or your employees, why is it okay to release them into the environment, or onto other people?

A BALANCING ACT

Another innovation from the Rocky Mountains Institute is the solutions economy, where the needs of the producer and the consumer are brought into balance. A solutions economy looks at what both the producer and the consumer really want, and then attempts to match the system to those needs.

Currently, most economies are based on limited-use products with built-in obsolescence. The reason fridges, washing machines, cars, and other consumer goods wear out is because they are designed to do so, thus forcing the consumer to buy new appliances, which is a waste of resources.

A more efficient system rewards both the consumer and the producer, with more durable goods that can be provided as leased services. Carrier, the world's largest maker of air conditioners, is a good example. They've teamed up with other companies dealing in building design and operation to provide new buildings with "air comfort," needing little or no air-conditioning. It makes good business sense for Carrier to provide cheap "air comfort services" without selling air conditioners because they are undercutting their competitors.

Similarly, Interface, a carpet company based in Atlanta in the United States, does not want to sell you carpet; it will, however, sell you a floor covering service. Instead of buying an entire floor space of carpet that needs to be replaced every ten years, they will carpet your office or home with carpet squares that they own. They will periodically inspect the carpet and replace only the parts that are worn or stained. Instead of throwing out a whole floor space of carpet that is only 10 percent worn, they replace just the sections of carpet that need attention. That's a 90 percent saving on carpet.

THE MOTIVE FOR CHANGE

The incentive to use the Earth's resources more wisely will not be a concern for the environment but rather a hard-core profit motive. It will be cheaper to use fewer resources, pollute less, and recycle more. Companies and institutions that adopt this approach to business will thus have a competitive edge over their rivals, which will result in greater profit that is not necessarily derived from greater production. Rival companies will then essentially be forced to adopt more environmentally sound practices or lose their market share.

The need to restructure economies and industries to be sustainable has impressed itself on governments. In 1996 the United States federal government started a working group as part of the Environmental Technology Task Force to conduct analysis and research on industrial ecology and materials and energy flows in the United States economy. Their findings and recommendations can be implemented in a variety of ways, including research, legislation, and consumer education.

There is obviously a long way to go before industry takes no more out of the environment than it puts back in—before it is truly sustainable. But the first few steps have been taken and a few enterprising and resourceful thinkers have lit the path to a profit-driven, environmentally sustainable industrial future.

SAND AND WATER *filters work to treat and recycle sewage water for later use by citrus farmers who irrigate their crops with it (above). The Sacramento Municipal Utility District in the United States (left) pays customers $100 to recycle their old refrigerators and gives $100 rebate on energy efficient ones.*

Digitizing the Senses

What's the difference between a computer and a pineapple?

As digital technology gears up to deliver true-to-life textures,

aromas, and flavors over the Internet, it may be difficult to tell.

We have five senses with which we perceive, understand, and delight in the world around us. But the growing use of computers has seen three of these senses—touch, taste, and smell—pretty much left out in the cold. Very recent advances in digital technology are now attempting to incorporate these forgotten senses into the digital world. Software and hardware developers are finding ways to bring the world inside your computer to life, with all its rough surfaces, diverse aromas, and tangy taste sensations, to create a total sensory experience.

VIRTUAL NOSE

One of the earliest efforts to digitize a human sense was the electronic nose. Artificial detectors of odors have been in existence since the 1950s, but only came into their own recently with the development of chip technology and pattern-recognition techniques. Now, electronic noses are being designed which can sniff out diseases, drugs, and even landmines.

One of the most sophisticated electronic noses has been patented by Cyrano Sciences in California, United States. This e-nose has 32 sensors made up of different polymers embedded with a conductive material. The polymers swell at different rates and change their conductivity when they encounter an odor. This conductivity change is recorded as a "smell print," which can be stored in a database.

E-noses don't tire like a human nose, and they willingly carry out unpleasant jobs that were previously done by medical staff, such as smelling urine samples for the presence of disease-causing microorganisms. E-noses are also being refined by forensic scientists to help assess the time of death of corpses.

TOUCH CAN BE A PLEASURE *both to give and to receive. Everyone has approximately 100 touch receptors on each of their fingertips alone. Of the 20 types of nerve endings covering the body, touch receptors are among the most common.*

TOUCHY-FEELY MOUSE

Modern computer graphics are so realistic you can almost reach out and touch them. But technology recently developed by Californian-based Immersion Corporation means you *can* actually feel what you see through an ordinary computer mouse. Run it across an image of a tennis racquet, and it "bumps" across the strings. Then move it over a picture of an ice cube and it suddenly seems to glide more smoothly. The TouchSense technology is derived from computer games, where, for example, the steering wheel shakes if you crash into something. The mouse works through a small motor concealed inside which feeds back sensations to the hand that guides it. As well as the obvious retail opportunities, surface textures on the Internet will give users a far more vivid experience of the world. On-line museums, for instance, may use the technology to show off exhibits which can be "touched" as well as seen.

THIS E-NOSE *(left), developed and used by NASA, is trained to detect changes in humidity and certain contaminants. E-nose technology will also enable us to capture and store the "smell print" of anything, from chocolate to the bouquet of a favorite wine, and even the perfume of a harvest of fresh roses (below).*

In future the e-nose may find its way into domestic life—inside our microwave ovens to smell when food is cooked, for instance, and even onto dairy farms to detect which cows are in heat and ready for artificial insemination.

CLICK AND SNIFF

Most exciting of all is the potential for the Internet to deliver real smells to the user. Imagine running your mouse across an image of a green apple and getting a whiff of its fresh smell, or playing a computer game set in an old castle and being able to smell the moldy walls you touch as you grope your way out.

Stimulating one's sense of smell is an effective way of accessing cherished memories and powerful emotions. Creating smells online would give advertisers, web developers, and moviemakers an enormous opportunity to influence their audience at a more subconscious level. Also, online shopping with digital scent technology will be a total sensory experience, developers say, allowing you to sample the bouquet of wines, compare the aroma of coffee from different countries, or indulge in the smell of fine Italian leather before purchasing.

Californian-based DigiScents is developing a scent delivery system for the home computer, rather like a PC printer. Dubbed iSmell, it contains 128 cartridges of scent compounds, all of which are different. These compounds can be mixed together rather like the cyan, magenta, and yellow inks used in printing, to create literally millions of smells. Scents can be tailor-made for individual use, but also saved as files

which could be e-mailed to friends. Web developers wanting to scent-enable their site will pay a licensing fee for the smell index, which describes the exact proportion of scent primaries for a particular odor.

As well as websites, scent technologies are likely to be incorporated into computer games, and future DVD movies may well be accompanied by a scent track as well as a sound track. Other applications might include online aromatherapy treatments, scents for music videos, and virtual cards smelling of roses, chocolate cake, or whatever you want. Internet sex sites will also no doubt want to use the technology to generate pheromones to enhance their product.

SMELLOVISION

The idea of matching smells to vision and sound is not a new one. Filmmaker John Waters experimented with scented movies, a genre he called Odorama, in his 1981 cult classic, *Polyester.* Cinema-goers were given a card with "scratch and sniff" panels to use at appropriate points during the movie. It was an intentional gimmick, but developers of scent technology today want to be taken far more seriously.

Watching the World's Weather

By the end of the 21st century, the world's population will

have access to weather predictions of unprecedented accuracy.

Since the days of primitive cave dwellers, the human race has been watching the weather—trying to anticipate its vagaries to make life safer and more comfortable. Weather watching has come a long way from the crude drawings on the walls of caves, but it has much further to go. Early 20th century weather forecasters struggled to provide even a rough forecast of the next day's weather. Forecasters today regularly make accurate predictions up to a week ahead. But in the 21st century, jokes about weather forecasts will become confined to the pages of history.

WEATHER IN HISTORY

The earliest systematic record of the weather dates back to the establishment of a weather station at the London Observatory in 1659. Records of temperature, originally using instruments crafted by the top artisans of the day, have been meticulously kept at this one

location ever since. Other weather stations progressively appeared in settlements around the world, but the great breakthrough in our attempts to predict the weather came with the invention of the Morse code by Samuel Morse in the 1830s. For the first time, weather observations from widely separated locations could be gathered in near real-time, enabling early forecasters to begin to observe basic patterns in the weather. During the 20th century these networks expanded across all oceans and continents, with ships, drifting buoys, and commercial aircraft transmitting their weather data to centers around the globe, freely exchanging information for the benefit of all.

The era of human weather observers is now rapidly drawing to a close as automatic weather stations proliferate across continents and remote islands, and transmit their data via high-speed communications. Satellites, all-sky and video cameras, ultrasonic and laser sensors will become commonplace in coming decades, so that the human role in weather observation is expected to transform into one of maintenance and calibration only.

THE WEATHER–WATCHING ARSENAL

The warring nature of the human race has spurred many technological developments that paradoxically prove to have lasting peaceful benefits. The advent of aerial warfare during World War II triggered the invention of compact instruments to measure temperature, wind, and humidity to high altitudes above Earth's surface. These instruments were borne aloft by balloons filled with hydrogen or helium, a technique still widely used today.

When radar built to protect cities from aerial bombardment was found capable of mapping out areas of rain or snow, the weather-watch radar was born. Such radar have grown in sophistication, with the newest having much greater sensitivity and resolution, along with Doppler wind processing. The Doppler technique measures the frequency shift of particles suspended in the air, and of rain, to determine whether these features are moving toward or away from the radar. The United States has

OBSERVERS *in the 1940s measure wind velocity by tracking a pilot balloon. Such hands-on activity is becoming a thing of the past, with calibration and maintenance of automatic weather stations taking over as key tasks for meteorologists.*

A METEOROLOGIST *services equipment at the University of Wisconsin Automatic Weather Station in Halley Bay, Antarctica (above). The image of Hurricane Florence (right), as seen from the space shuttle Atlantis on November 14, 1994, shows a typical hurricane system comprising a large-scale spiraling line of clouds over an intense weather front.*

by far the most comprehensive network of weather-watch radar, with NEXRAD radar providing almost total coverage of the United States mainland. In the coming decades, this type of radar will be installed in other countries. Dual polarization techniques, which identify hail suspended inside clouds, will also move from research into operational use.

As important as the development of radar, was the invention of the weather satellite,

LOW-EARTH ORBIT SATELLITES—ADEOS

Innovation in science has led to the development of sophisticated low-Earth orbit satellites carrying delicate sensors that can measure a wide range of environmental parameters. The ADEOS series of satellites, a joint Japan-United States mission, will measure everything from the ozone content of the atmosphere and ocean color (chlorophyll) to oceanic temperatures, winds, waves, and topography. Other sensors detect things as disparate as soil moisture, atmospheric moisture content, floods, and ocean currents, all of which are necessary components for a more complete understanding of our weather systems.

which provided the quantum leap in our ability to predict the world's weather. The first weather satellites showed patterns in the clouds that astounded forecasters of the early 1960s. Of the two types of satellites used for weather-watching, the earliest type launched were low-Earth orbit satellites, typically orbiting at an altitude of about 500 miles (800 km). These satellites follow a pole-to-pole orbit in order to provide very high resolution images and vertical temperature profiles of the atmosphere.

The second type were high altitude satellites that sit about 22,230 miles (35,800 km) above the equator and orbit the Earth once every 24 hours in order to achieve a constant view of it. Today, geosynchronous satellites (see box feature overleaf) provide minute-by-minute updates of storms as they develop over the continental United States. This will

THE AEROSONDE *(left) is a small robotic aircraft developed especially for long-range reconnaissance over oceanic and remote areas, and in harsh conditions. It is being deployed to fill chronic gaps in the global upper-air sounding network. The radar image (above) is of the Weddell Sea, Antarctica. Taken from the space shuttle Endeavour, it shows two large eddies just visible at the northernmost edge of the pack ice.*

cially strengthened military aircraft. They can also track other weather systems across vast uninhabited areas, providing details of incipient storm development well before such storms reach heavily populated regions.

undoubtedly evolve into a global capability by the middle of the 21st century thanks to widespread international cooperation between the world's satellite-operating nations. Their satellite programs are being coordinated to maximize the coverage and capabilities of the weather satellites to be launched by each participating nation.

Another fascinating new tool in the weather-watching arsenal is an experimental, remote-controlled aircraft called an aerosonde, capable of autonomous operation and filled with meteorological sensors. With a 10-foot (3-m) wingspan, this aircraft can fly into hurricanes while they are still hundreds of miles out to sea, accurately tracking their movements and intensity changes, and replacing expensive surveillance currently only performed by spe-

FUTURE FORECASTS

Globally coordinated research projects are greatly assisting in our understanding of the world's most powerful and severe weather systems, from monsoons to hurricanes to the mighty twins, El Niño and La Niña. The Tropical Ocean Global Atmosphere (TOGA) and Tropical Rainfall Measurement Mission (TRMM) research programs initiated in the 1990s are improving our understanding of the tropical systems that are the powerhouse of the world's weather. Oceanic research projects such as the World Ocean Circulation Experiment (WOCE) will lead to improved longer-range predictions as it becomes clear that the oceans greatly influence the large-scale weather circulations of the world.

The world's fastest supercomputers are already devoted to simulating constantly evolving weather patterns, yet the available computing power still can't provide sufficient detail to predict the weather in either the very short or the long range. The massive volume of high-frequency satellite, radar, and surface observations that contain the clues to future weather patterns already swamp the forecaster of today. The next generation satellites will increase this by several orders of magnitude. More advanced computer simulations of the weather, which use all this data and include real-time interactions between the atmosphere, oceans, land, and ice sheets, are under development. The improved simulations will cover not only the weather elements that we are familiar with (wind, rain, temperature, and so forth), but also improved representations of features such as clouds, thunderstorms, snow, large waves, and evaporation that are crudely represented at present. These predictions will provide detailed forecasts that were a dream when the first weather simulations were being conducted.

In 1922 a British mathematician and scientist, Lewis Fry Richardson, took a year to apply mathematical techniques to produce a crude 24-hour weather forecast, a forerunner of the numerical prediction techniques of today. Ensemble modeling techniques, which use multiple computer runs to give a range of possible outcomes for the weather for a given time, identify the chaotic nature of the atmosphere. Such techniques will continue to be refined, providing predictions whose reliability will be known ahead of time. Statistical techniques that utilize the increasingly detailed surface observations, and more comprehensive historical surface and satellite observations, will underpin weather predictions for the seasons, and even the year, ahead.

The complex pattern recognition processes that the best forecasters currently use to identify the early stages of flash floods, severe squalls, tornadoes, and large hail from three-dimensional radar data are being tackled in another fascinating manner. Neural network and artificial intelligence techniques that replicate the logical processes of the human brain are being applied to more reliably identify the transient patterns of severe weather events. The power of computer-based techniques lies in the volume of data that can be processed repeatedly without risk of a lapse in concentration or any other limitation to which humans are vulnerable.

WEATHER SATELLITES

The first weather satellite was an awkward-looking contraption named TIROS (Television Infrared Observation Satellite), launched by the United States in April 1960. It followed a low-Earth orbit and took black and white infrared pictures of the Earth. More advanced polar orbiting satellites followed, with the Europeans, Russians, and later Chinese and Japanese launching their own. The next milestone occurred with the launch of the first weather satellite that orbited directly above the Earth's equator in 1966. This magical geosynchronous orbit is ideal for monitoring severe weather since it permits the continuous tracking of the life cycle of ever-changing destructive storms. Weather satellites in this type of orbit have increased in number and complexity. Today, there are seven such geosynchronous satellites, operated by the United States (GOES East and West), Russia (GOMS), China (Feng-Yun), Japan (GMS), India (INSAT), and the European Union (METEOSAT). The Meteostat Second Generation satellites (artwork, above) will provide clearer, more realistic images that will allow meteorologists to provide more accurate medium and short-term forecasts.

Natural Disasters

Extreme and violent acts of nature now threaten more people

around the world than at any other time in human history.

As natural disasters wreaked death and destruction across the globe on a massive scale during the late 20th century, proponents of a new millennium apocalypse must have believed their prophecy was coming true. The world, of course, didn't end as the old century departed, but the new one arrived with scientific forecasts of an even bleaker outlook for natural disasters. During the decades to come, hazards such as floods, droughts, earthquakes, landslides and hurricanes threaten to cause an unprecedented loss of human life, property damage, and social and economic upheaval. And there is nothing supernatural behind the reasons why.

Hazardous Population Growth

Several features of human population growth give natural disaster experts particular cause for concern. The first is the massive global trend toward urban living. Already, half the planet's six billion people live in cities and they are expected to be joined by another five billion by 2050. As a result, two-thirds of the world's population is likely to be crammed into urban areas within the next 50 years.

Natural disasters have a far more catastrophic effect when they hit heavily built-up areas. And so the shift towards high-density living is statistically putting greater numbers of lives at risk. The problem is exacerbated by the fact that much of the urban population increase is occurring in the world's poorer areas. Already, over 95 percent of natural disaster deaths occur in developing nations where poverty precludes the early warning systems, urban planning, and design and construction methods necessary to minimize the impact of natural disasters.

In another worrying trend, a greater number of people than ever before live in Earth's most disaster-prone areas, and still more are moving there. The reason is that the geography and climate of these areas often makes them largely pleasant places to live. New retirement settlements, for example, continue to spring up in hurricane-vulnerable areas of the United States' "Sun Belt." Elsewhere, the number of people living at risk from volcanic eruptions is at least 500 million and rising. A further billion live in shanty towns perched haphazardly atop seismic fault lines. And 40 of the world's 50 fastest-growing cities are located in earthquake zones.

ONGOING MONITORING *of the Philippines' Mount Pinatubo by volcanologists (above) has already saved thousands of lives and could do so again in the 21st century. Less easy to monitor is high rainfall like that experienced in south-west Bangladesh in September 2000, which left more than 800 dead and millions homeless. Thousands tried to escape rising flood waters by fleeing along a national highway to West Bengal (opposite).*

HUMAN IMPACT

There is evidence to suggest that, as these population trends put more and more lives at risk from natural hazards, agricultural and industrial activities are boosting the number and intensity of weather-related disasters. Deforestation, poor water management, and over-cultivation are blamed, for example, for an increase in floods, droughts, and landslides. Climate change scientists predict that global warming could lead to more of the same, as well as an increase in violent storms such as hurricanes.

British development agency Tearfund nominates climate change as a major cause of a four-fold increase in the annual number of "extreme natural disaster events" in the last 50 years of the 20th century. And some experts suggest the planet may experience a "mega-drought" made so severe by global warming it could lead to a natural disaster bigger than anything seen during the 20th century.

There is no evidence that geology-related natural hazards such as earthquakes, volcanic eruptions, or tsunamis ("tidal waves" caused by seismic activity) have increased or will do so as a result of human activity. However,

GLOBAL RESPONSE

Confronted by a growing mountain of evidence that risks posed by natural disasters were escalating, the United Nations took action in 1987 and designated the final 10 years of the 20th century as the International Decade for Natural Disaster Reduction. The initiative focused world attention on the magnitude of the problem and stimulated significant advances in natural disaster forecasting and mitigation. The global human and economic toll of natural disasters continued in that decade, however, to spiral upward. And so, with the close of the decade, the United Nations created the Geneva-based International Strategy for Disaster Reduction to continue coordinating programs worldwide to reduce the impact of natural disasters in the 21st century.

many of our practices could exacerbate the impact of these hazards. Coastal communities living in some parts of tropical Asia, for example, may now be at greater risk from tsunami damage due to the clearing of mangrove thickets for port facilities and shrimp farms.

EARLY WARNINGS SAVE LIVES

There are no technologies available that can alter the paths of killer hurricanes, stop earthquakes, or contain violent volcanic eruptions, and we're unlikely to see anything of the sort during the 21st century. Over the past decade,

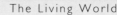

THIS 3-D PERSPECTIVE *of Mount Oyama (above), which erupted violently in 2000 on the Japanese island of Miyake-Jima, was generated using topographic data from the Shuttle Radar Topography Mission. Valuable for predicting severe weather is the Doppler radar (left), which measures the speed and direction of moving weather phenomena such as raindrops.*

however, scientists and engineers have substantially improved their capabilities to predict many natural hazards and prepare communities.

One of the most spectacular saves of the 20th century occurred on June 15, 1991, in the Philippines, when Mount Pinatubo exploded after "sleeping" for over 500 years. The homes of over one million people were located within 30 miles (48 km) of the explosion—the 20th century's largest volcanic eruption to affect a heavily populated area. Almost 900 people were killed—but that figure was a lot lower than it could have been. Scientists from the Philippine Institute of Volcanology and Seismology and the U.S. Geological Survey predicted the eruption in time for a large-scale evacuation. According to conservative estimates, at least 5,000 lives and $250 million worth of property were saved.

Improvements in natural disaster early warning systems have occurred largely due to the increase in information technologies, telecommunications, and telemetry. In particular, more recent developments in computer modeling, remote sensing, and the Internet have provided scientists with sophisticated tools with which to gather information about natural hazard risks, analyze that information, and ultimately make predictions and disseminate warnings. With the aid of Doppler radar, satellites, and specially equipped aircraft, meteorologists can now hunt down and predict the progress of more dangerous, fast-moving weather events such as hail and tropical storm formation. And, by using technologies such as satellite remote sensing, it's possible to construct hazard maps of risk areas to guide appropriate development and land-use practices in hazard-prone areas.

The sense of global cooperation that pervades the disaster prevention community is extraordinary—from seismologists, engineers, hydrographers, climatologists, geologists, and

meteorologists, to social scientists and aid workers. This global spirit of cooperation has also been a feature of the majority of scientific developments in the area. It is exemplified by the system of disaster monitoring and warning stations that have been extending their reach around the world in recent years.

Tsunami Warning Centers throughout the Pacific, for example, are charged with detecting, locating and measuring the magnitude of earthquakes that may cause giant "tidal waves." A Global Disaster Information Network was established in the late 1990s. There is a Global Fire Monitoring Center based in Germany and the Geneva-based World Meteorological Organization oversees a Global System of Meteorological Satellites and Drought Monitoring Centers.

SUPERHAZARDS

There are two types of natural hazard with potential to create a global-scale disaster, and scientists know very little about either. A supervolcano eruption is one. An impact by a large—greater than half a mile (1 km) in diameter—asteroid or comet, is the other. At least eight supervolcanoes lie hidden below the Earth's surface, including one six miles (10 km) beneath North America's Yellowstone National Park. According to volcanologists' calculations, this blows every 600,000 years, making the next major eruption already 40,000 years overdue. Supervolcano eruptions are far more massive and destructive than eruptions from ordinary volcanoes.

A DECADE OF TRAGEDY

The 1990s were notable for a large number of "extreme" natural disasters, which caused mass death and destruction. These were just a few of those tragic events: In Iran, in 1990, an earthquake-related landslide killed an estimated 50,000 people. The following year a cyclone and associated flooding claimed 140,000 lives in Bangladesh, and Typhoon Thelma killed more than 6,000 in the Philippines. The 1995 Kobe earthquake, in Japan, left more than 5,000 dead and a damages bill in excess of $100 billion. Hurricane Mitch (above), the Atlantic's most powerful storm in 200 years, tore through Central America in 1998, leaving at least 11,000 dead. A tsunami killed 3,000 in Papua New Guinea in the same year, while flooding of the Yangtze River, in China, claimed over 4,000 lives. In 1999, earthquakes killed over 14,000 in Turkey and 2,000 in Taiwan.

Climatologists and geologists believe that when the supervolcano Toba, in Sumatra, erupted 75,000 years ago it altered the Earth's climate and, according to recent genetic studies, may have pushed the human species to the brink of extinction. A large meteor colliding with Earth would have similarly devastating effects, although the likelihood of such an impact is extremely small. However, it has happened before (wiping out the dinosaurs and 70 percent of all life some 65 million years ago) and it will happen again. Because of its potentially catastrophic effects, concerned scientists began developing programs to identify and track the movements of large meteors and asteroids.

THE MASSIVE LAKE TOBA, on the island of Sumatra, was created 75,000 years ago by a huge volcanic eruption unlike anything since seen on Earth.

Biodiversity

Our fate rests no longer on whether two nuclear powers start

a war, but on how well we protect the Earth's biodiversity

in the next few decades.

Biodiversity is the sum total of all living things on the planet, and all the genes they contain. Recent research shows that biodiversity is essential to maintaining the health of all ecosystems on Earth. Remove one or several species, and you create changes that reverberate through the whole system.

This is likely to have happened in the Southern Ocean when the blue whale was hunted almost to extinction during the first half of the 20th century. Squid, seals, and seabirds were the blue whale's main competitors for shrimp-like krill, upon which they all feed. Once the whales were gone, the populations of these other predators appear to have exploded, effectively taking over the ecological niche that the whales had filled. Because of this, there now may not be enough krill for the blue whale to reach its former numbers ever again, despite a ban on the fishing of them since the 1960s.

SCIENTISTS MEASURE THE WINGSPAN *of a Wandering Albatross on South Georgia Island in the South Atlantic Ocean (above). The Wandering Albatross is just one of the thousands of animal species teetering on the edge of extinction.*

Biodiversity also gives ecosystems a better chance to cope with changes, such as global warming, fire, or drought. Just as the most stable investment portfolio is one that is spread across several companies, ecosystems can better withstand environmental fluctuations if they are made up of species with different tolerances to temperature, rainfall, and nutrients.

UNDISCOVERED RICHES

Ongoing discoveries of new species will be a major area of biodiversity research in the near future. Scientists have identified 1.75 million species, but estimate that there are probably over 12 million species still to be found. For

BIODIVERSITY HELPS THINGS GROW

Researchers in the United States looked at how diversity affects growth. They set up farming plots with different numbers of species, then looked at overall productivity (the total biomass produced over time.) The results were startling. A plot with one species turned out to be only half as productive as the same size plot with up to 32 species. With each halving of the number of species in a plot, up to 20 percent of growth was lost.

the five previous mass extinctions on Earth put together. The International Biodiversity Observation Year Advisory Board says one-third or more of all species, many not yet known to science, are threatened in the next few decades. The World Conservation Union (WCU) estimates that one in four species of mammal and one in eight species of bird are at risk of extinction—as are 25 percent of reptiles, 20 percent of amphibians, and 30 percent of fish. In the last four years of the 20th century, the WCU added an extra 230 species to its list of those threatened with extinction. These include the Wandering Albatross with its 11-foot (3.4 m) wingspan, and the sturgeon fish from the Caspian Sea, famous for its eggs, or caviar.

99 percent of known species, there is little information on their distribution or abundance, or what role they might play in renewing soil fertility, decomposing waste, or cleaning water.

New discoveries are being made constantly. Most recently, a totally new kind of microbe was found living 1½ miles (2 km) below the surface of Earth. Another new discovery—a nanobacterium—was found living inside human kidney cells. Similarly, new species are continually being found, many of which may have medicinal value or play a vital role in maintaining ecosystem health. Recent research in South American rainforests has netted hundreds of species of insects previously unknown to science. Other species, which were thought to be extinct, are being rediscovered in remote areas. For example, a giant stick insect recently found near Lord Howe Island, Australia, was believed to have become extinct 80 years ago.

HOW MUCH ARE WE LOSING?

Human activity has led to extinction rates up to 1,000 times what they were before humans walked the Earth. If current land use and other damaging activities continue unchanged, we will lose as much biodiversity this century as

RAINFORESTS *are constantly yielding many previously unknown species, such as the flat-headed katydid (Lirometopum coronatum, above). To collect insects at night, scientists in Chihuahua, Mexico, illuminate a tarpaulin, then gather some of the insects that are attracted to it (left). At present extinction rates, it is feared that some species will disappear before they have been discovered and cataloged.*

MADAGASCAR *occupies only 1.9 percent of African land area, but has more orchids than the entire African mainland and is home to about 25 percent of all African plants, all of the continent's lemurs, and most of its reptiles and amphibians. Its Tsingy de Bemaraha Strict Nature Reserve (above) is the habitat for rare and endangered lemurs and birds, but many of the island's other areas of biological wealth are threatened by slash-and-burn agriculture, timber exploitation, uncontrolled ranching of livestock, and hunting. Conservation International has identified 25 biodiversity hotspots (below)—biologically rich areas that are under the greatest threat of destruction and in which conservation will have the greatest global impact.*

WHAT CAN BE DONE?

The five biggest contributors to biodiversity loss in the next 100 years are expected to be atmospheric carbon dioxide, climate, the introduction of exotic species, nitrogen deposition by industry, and land use. Researchers have recently calculated the contribution of each of these to biodiversity losses in different habitats, from alpine ranges through to lakes and rivers.

Biodiversity in Mediterranean and grassland ecosystems will be the most affected by all five contributors. Land use by humans poses the

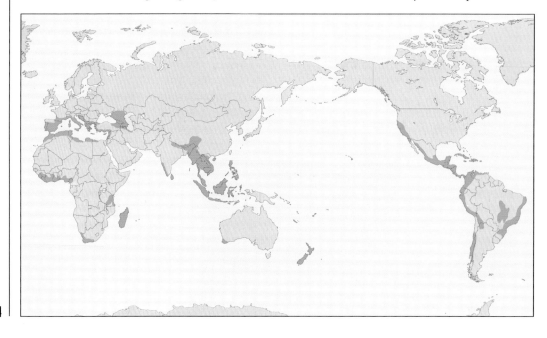

THE PILL CAME FROM A YAM

Biodiversity has supplied the raw materials for thousands of medicines and drugs that have altered the course of human society. The discovery of a humble yam from India some decades ago, for example, caused no less than a social revolution. The yam, *Dioscorea deltoidea*, was found to contain high concentrations of diosgenin, a cortico-steroid, in its tubers. Eventually this chemical was used to form the basis of contraceptive pills to control ovulation.

greatest threat to terrestrial ecosystems, and freshwater ecosystems are most vulnerable to invading exotic species. Climate change is the biggest threat to alpine communities, and nitrogen deposition near northern cities is the major risk to biodiversity in temperate zones.

Other research focuses on giving top priority to protecting "biodiversity hotspots"—areas that are both high in biodiversity and rich in species that live nowhere else. British ecologist Norman Myers, Harvard University's E. O. Wilson, and other experts have identified 25 such hotspots. These once covered 12 percent of the land surface of the planet, but have shrunk today to 1.4 percent (an area about three times the size of Texas). Even so, this tiny amount of land contains nearly half of all vascular (woody) plants and over one-third of all mammals, birds, reptiles, and amphibians. This concentrated biodiversity richness makes these hotspots an extremely high priority for future protection efforts. They include Madagascar, the Philippines, the Sunda Islands, Brazil and the Brazilian Atlantic Forests, the Caribbean, a southwesterly fragment of Australia, New Caledonia, and New Zealand.

Conservation International, the organization that undertook this research, says protecting biodiversity in the future will need to be done on a scale never before proposed. It will seek funds for a multibillion-dollar plan (provisionally called An Agenda to Defy Nature's End) from philanthropists and private individuals. Such schemes suggest that future biodiversity research may be driven more by private than public dollars, although some scientists argue that such plans will not work without cooperation from the governments involved.

Efforts are also being made to conserve what is left of the natural biodiversity on which humans rely directly for food. During the postwar Green Revolution, a great many traditional crop varieties disappeared from fields and orchards to be replaced by high-yielding hybrids. While these grew faster and bigger than the older varieties, the typical commercial hybrid lasts only 5 to 10 years before it needs genetic infusions from the older strains to retain its vigor. But these older strains, and the wealth of biodiversity that they represent, are now being preserved with new urgency by the International Plant Genetic Resources Institute. The IPGRI is assembling a global network of seed banks that recently topped six million varieties. Already the bank has been called upon to resupply traditional varieties to regions where agriculture has been severely disrupted by war or famine, such as in Cambodia after the overthrow of the Khmer Rouge.

Researchers hope that identifying the risks to biodiversity will help focus resources where they are needed most before it is too late. But they warn that efforts to protect biodiversity will need an integrated approach by ecologists, climatologists, social scientists, and policy makers if what we have left is to be protected for future generations.

SCHOOLS SAVING BIODIVERSITY

Students at a school in Tasmania, Australia, are thinking globally and acting locally. They want to reestablish habitat for a threatened bird, the Swift Parrot (*Lathamus discolor*). The parrot relies on a particular tree species, the Tasmanian Blue Gum, but these are disappearing from the area. Students have begun growing seedlings of the tree around the school in the hope of establishing a future habitat for the parrot.

The Shifting Earth

Earthquakes are the single most dangerous geological hazard

humanity faces—can anything be done to predict them?

Every year it happens. The locale may vary—Japan, Turkey, Central America, or perhaps Italy—but the story is always the same. The ground shakes, buildings collapse, and people die. Every year sees about 18 earthquakes of magnitude 7.0 or stronger, and about 10 times that number between magnitudes 6 and 7. These seismic disturbances are large enough to cause severe shaking and extensive damage to most buildings. By any measure, they are killers.

The threat of earthquakes touches much of the world, thanks to the relentless power of plate tectonics. Earth's crust is broken into approximately two dozen plates whose edges slowly grind against each other (or which override their neighbors) under the force of churning rock movements in Earth's hot mantle. Seismic areas most often lie where plates adjoin, including Italy and Greece, Turkey, Central Asia and India, China, Japan, the mid-ocean ridges, and the entire western coasts of North and South America.

Fortunately, most large earthquakes occur in regions that are uninhabited or sparsely populated, but many do not. And if earthquakes are a normal part of living on a geologically active planet, is there any hope of predicting them? In the 21st century seismologists have less optimism on this score than they once did. But they are putting the latest techniques to work as they search for new ways to forecast when the solid earth will tremble.

HUNTING FOR PRECURSORS

On February 4, 1975, a magnitude 7.4 quake violently shook the north Chinese city of Hai-cheng, some 350 miles (550 km) northeast of Beijing. In the months before the quake, many people in the region had observed that water levels in wells had changed, and they saw a marked increase of restlessness in the animals. This peculiar behavior intensified during the first few days of February, until several small earthquakes on the morning of February 4th convinced local officials to issue an urgent earthquake warning. The quake struck that evening, but the warning meant that relatively few people were hurt, despite most of the city being severely damaged.

Eighteen months later, again in China, a different story unfolded. On July 28, 1976, a magnitude 7.8 quake struck the industrial city of Tangshan, east of Beijing. In about 10 seconds of shaking, the whole city was leveled. This time there were few advance warning signs, and no alert was given. The quake struck at around 4 a.m., catching most people in their beds. Over 255,000 people died.

To predict earthquakes, geologists need reliable precursors, seismological evidence that an earthquake is imminent. The list of candidates is long. It includes changes in rocks' electrical resistivity, anomalous animal behavior, tidal forces, slow deformation of the ground surface, and many others.

Outside China, evidence from animal behavior is not considered significant because no studies have been able to link specific behavior to geological activity. Changes of water levels in wells may be considered more useful, but variations occur routinely without seismic causes. Tidal stresses from the Sun and Moon sound intriguing, but they don't correlate closely with earthquake activity.

More useful in the prediction of quakes has been the concept of seismic gaps. Most quakes

SECTIONS OF THE JAPANESE *city of Kobe (above) were completely demolished when a powerful quake struck in 1995. Japan lies where one crustal plate is overriding another, and it is heavily seismic. Across the Pacific (left), offset hills show where one of the world's great fault lines—California's San Andreas fault— has shifted.*

occur on known faults, and if past activity has been somewhat regular, it may hint at when a given fault is "due." Also, the pattern of quakes along a lengthy fracture, such as the North Anatolian fault in Turkey, can show where parts of the fault are more active than others. This could indicate that a quiet section in between may be ready to slip.

PREDICTABLE PARKFIELD?

Yet even when many precursors are combined, there's no guaranteeing a prediction. In the United States, prediction efforts have centered mostly along the best-known and studied fault, California's San Andreas, and in particular on the tiny town of Parkfield between Bakersfield and Monterey. Parkfield experienced quakes of magnitude 5.5-6.5 in 1881, 1901, 1922, 1934, and 1966. Scientists, noting the approximate recurrences, wrapped the town in a geophysical

network of seismometers, tilt- and creepmeters, strain-gauges, and other sensory equipment.

From the wealth of data that was collected, the United States Geological Survey issued its first-ever earthquake prediction in April 1985. It said that Parkfield had a 95 percent chance of experiencing a magnitude 6 earthquake in January 1988, plus or minus four years. When the quake hadn't materialized by December 1992, the organization had to admit failure. Parkfield still hadn't experienced the predicted earthquake by the end of the 20th century—indicating, if nothing else, that reading precursors is tougher than scientists thought.

INTERPRETING SEISMIC SILENCE

People living along active fault lines can expect earthquakes from time to time. Their best protection is to avoid areas that suffer severe shaking, and to construct buildings that are quake-resistant. More dangerous are places that are geologically at risk, but where there have been few or no major quakes in historical times. One such place is the northwestern United States, where plate tectonics is forcing a piece of Pacific Ocean floor called the Juan de Fuca plate beneath the edge of the continent at about 1.6 inches (40 mm) a year.

Geologists call these regions convergence zones, and note that they fall into two types. One has very small quakes almost daily as the descending plate slips easily into the upper

LASER BEAMS FROM RANGEFINDERS *fan out near Parkfield, California (above), in a seismic test that is part of the effort to monitor activity along the San Andreas fault. Despite a great deal of data-gathering, however, reliable predictions of earthquakes have proven to be difficult..*

mantle. In the other, the plate is warmer and more buoyant and disappears less smoothly. This type of zone features large earthquakes at infrequent intervals. The Juan de Fuca plate is of the second kind, and in its seismic region lie several major cities, including Vancouver, Seattle, and Portland, all of which are at serious risk of a major earthquake.

Seattle, for instance, sits atop a thick pile of sediment filling a rocky basin whose floor is crisscrossed by faults that seismologists have just begun to map in detail. (Knowing where the faults run is essential for disaster planning: The 1994 Northridge quake occurred on a Los Angeles-area fault unknown before the quake.) The magnitude 6.7 quake that struck in February 2001 did only minor damage, mainly to older buildings. In a future quake, however, the area might not be so lucky. Experience with earthquakes elsewhere shows that sedimentary infill amplifies seismic motions, in much the same way as jelly in a bowl sways wildly when the bowl is struck.

Seismic silence prevails elsewhere in the United States, including regions that are generally believed safe from major disturbances. For example, the central United States appears seismically quiet—yet the region around the Mississippi River town of New Madrid, Missouri, was literally shaken to bits by powerful quakes in 1811 and 1812. The few buildings existing at the time collapsed, groundwaves about 2 feet (60 cm) high were seen, and nearby Reelfoot Lake was created in Tennessee. The biggest quake woke President Madison in Washington and rang church bells in Boston hundreds of miles (kms) away. The quakes occurred along faults hidden under layers of river sediments.

Back then the region was thinly populated, but a repeat of the quakes today would kill hundreds of thousands and perhaps destroy Memphis, Tennessee, and St Louis, Missouri. Human life apart, the cost would run into untold billions of dollars.

CHANCE OF QUAKES: 90 PERCENT

Instead of predicting when a quake may strike a particular city, seismologists now tend to focus largely on forecasting in a more general way—by estimating the likelihood that a quake of some particular strength will occur within a given timeframe over a broad area. The forecast is not as detailed as people would like, but it's the best that present science can do.

Except for China, most places lack a long record of historical quakes, and the science of seismology itself is still very young. The modern seismograph wasn't even invented until the 1890s, and the ability to monitor the entire planet's seismic activity dates from only around 1950. Even now, the global network of instruments has many sizeable gaps.

New techniques of instrumentation, however, hold the promise of much better information. For example, geoscientists can image a region with synthetic aperture radar from high-flying aircraft and satellites in orbit. The result

is a map of elevation changes measured in inches (a few cm) over large areas. The build-up (and release) of stress in the ground can be accurately monitored—as can stress patterns following a quake, which often changes the forces on neighboring fault systems. This may increase (or decrease) their chances of slipping.

THE GLOBAL VIEW

Where earthquake forecasting will be a century from now is unguessable. Giant supercomputers digesting vast amounts of seismic data should be able to produce better, more detailed forecasts in the decades to come. Meanwhile, in the most seismically prone regions, dense networks of instruments can track the slow build-up of deformation along a fault. Then if the fault lets go, they can determine, essentially in real-time, how the quake-in-progress is moving and how violent it is. This automated analysis allows a kind of "prediction" on very short time-scales. When the quake begins, a warning can be flashed to critical places farther away, such as disaster-response headquarters, power stations, and rail- and transport-control centers.

With a degree of luck, the warning may arrive many seconds to a minute or two ahead of the most damaging waves, and give authorities a small chance to get emergency measures under way. Mexico City, for example, lies about 50 seconds away (in quake-travel time) from the country's most seismic region along the Pacific coast. And in Japan, the high-speed rail system uses instrument networks to start automatically braking its bullet trains when seismic trouble develops.

Aside from maintaining a global seismograph system to survey activity, perhaps the most useful thing geoscientists can do is develop detailed maps of where seismic hazards are greatest, and help engineers design structures better equipped to survive strong shaking.

Earthquakes, volcanoes, and the slow grind of plate tectonics tell us that nature hasn't yet finished making planet Earth. The job of seismology is to tell us more about our planet, and give us the knowledge we need to avoid being victims of its impersonal forces.

USING THE SATELLITE-BASED *Global Positioning System (GPS) (left), earth scientists can monitor the slow creep of the ground in response to seismic forces.*

Extinctions

A growing number of scientists believe that

the Earth is facing a period of mass extinction

unlike anything it has experienced before.

About 65 million years ago, life was decimated during one of the planet's biggest extinction events ever. More than three-quarters of all species alive at the time, including the last dinosaurs, disappeared forever. It was, the fossil record indicates, the fifth such "mass extinction" to occur since life began four billion years ago.

Now, many biologists, ecologists, and pale-ontologists believe that a similar catastrophe is already under way—the sixth mass extinction. They predict that, based on current trends, as many as half of the plant and animal species with which we currently share the planet could disappear during the 21st century.

Extinction is an inevitable part of evolution. Around 99 percent of all species that have ever evolved have become extinct, most of them after one million to 10 million years of exis-tence. Scientists believe that, under normal conditions, the planet should be losing approx-imately one species every four years. According to conservative estimates, however, we are al-ready losing 17,000 species a year, and the vast

majority of those are lost before we have had a chance to scientifically record their existence.

Several features set the sixth extinction apart from the previous five. To begin with, there is the pace of it. All past mass extinctions have occurred over hundreds of thousands, and sometimes millions, of years. But from just 1950 to 2000, ecologists estimate that Earth lost hundreds of thousands of species.

Scientists agree that past mass extinctions were linked to some form of natural phenom-enon. But they blame humans for the current crisis, listing the top causes of modern extinc-tions as hunting and poaching, and the clearing and fragmentation of habitats.

Could the human race be ultimately wiped out by the current mass extinction? Quite possibly. As Stanford University biologist Paul Erlich puts it, "In pushing other species to extinction, humanity is busy sawing off the limb on which it is perched."

The good news is that many scientists believe we have the means, and just enough time available, to significantly slow, and pos-sibly even halt, the extinction drama that is threatening to unfold in the 21st century.

WHEN WORLDS COLLIDE

As previous mass extinctions suggest, there are other ways in which the human species could ultimately meet its demise. An object from space the size of the "dinosaur-killer" asteroid widely believed to have caused the last mass extinction collides with Earth, on average, once every 100 million years. If one hit Earth today, its effect on the planet's atmosphere would probably wipe out humans and many other species. Smaller objects, the likes of which (statistics say) hit Earth every 20,000 to 200,000 years, could inflict widespread death and destruction on human civilizations.

Astronomers predict an asteroid over half a mile (1 km) wide and known as 1999 AN10 will pass close to the Earth in 2027. They are concerned that it could come near enough for Earth's gravitational pull to alter the asteroid's

THE TARBOSAURUS BATAAR *inhabited Central Asia about 75 million years ago. It is widely believed that the last dinosaurs were destroyed by the Chicxulub asteroid collision in the Yucatan peninsula region of Mexico 65 million years ago.*

HABITAT CLEARING *(above) is thought to be a major contributing factors to the current mass extinction. The World Resources Institute estimates more than 50 million acres (20 million hectares) of tropical forest face a similar fate each year.*

trajectory in a way that could put us at a very small risk of a direct hit in 2039. 1999 AN10 is just one reason astronomers around the world are keen to keep watch on these "near-Earth objects" (NEOs) which hurtle through space. Comet Shoemaker-Levy 9's cataclysmic impact with Jupiter in 1994 is another. And then there is the extraordinary Tunguska fireball that exploded in 1908 over an uninhabited part of Siberia with the force of 15 million tons (15.2 million tonnes) of TNT.

What, if anything, could scientists do if they detected a dangerous NEO hurtling toward us? A number of schemes, all extremely risky, have been proposed. For example, a rocket armed with nuclear warheads could be aimed at the NEO with the intent of either destroying it or knocking it onto a safer course.

Back on Earth, the risks of extinction posed by the same warheads during the next century are probably much greater than those from an NEO. There have reportedly been at least three occasions during the past two decades when the United States and the Soviet Union came close to launching nuclear warheads in response to false alarms. In that time there have also been numerous scientific studies showing

IN THE 1920s, *the Barringer Meteorite Crater in the Arizona Desert in the United States was the first site confirmed by scientists as being caused by a large object from space. At least 150 similar sites have since been identified on Earth.*

that nuclear war (even on a small scale) and the ensuing "nuclear winter" would push the human species to the brink of extinction. While the threat of a nuclear war abated during the late 20th century due to improvements in international relations and diplomacy, concern remains that many warheads remain primed for release and may be launched accidentally, in response to a false alarm.

Bioweaponry

Compared with other potential weapons of mass destruction,

biological weapons are cheaper to produce, easier to conceal,

and have the weakest international prohibition regime.

The world's capabilities for biological warfare and terrorism have never been more frightening. Technological advances during the closing decades of the 20th century served to increase the ease with which bioweapons can be developed and the destruction they can cause. The accelerating pace of global travel, for example, provides the opportunity for a much more rapid and extensive spread of deliberate biological infections than ever before. And the Internet can furnish unidentifiable suppliers of the raw materials, equipment, and technological know-how used in bioweaponry production.

This century, science's unwitting "gift" to bioweaponry is biotechnology. Although this area promises to underpin a great many positive scientific developments during the coming decades, it could also provide some of humanity's darkest moments.

Bioweapons are designed to kill or debilitate people, livestock, or crops. Biological agents with the potential to be turned into weapons include anthrax, plague, tularemia bacteria, smallpox and Ebola viruses, and plant-disease fungi such as potato blight. Other harmful biological substances with the same potential include botulinum toxin and staphylococcal

enterotoxin, both from bacteria, and ricin from the castor bean plant. Biotechnology could be used to improve the weaponry potential of many of these. For example, antibiotic-resistant bacteria, or viruses genetically engineered to make them immune to vaccines, could be produced. According to widespread reports during the 1990s, the Soviet Union developed a strain of antibiotic-resistant plague as part of its Cold War biological weapons program. Bioweapons that are targeted at specific population groups who share common genetic traits, such as sometimes occur in particular ethnic groups, are another disturbing possibility.

IMPOSSIBLE TO VANQUISH
Infectious agents have been used as weapons for centuries, and during the Cold War dedicated bioweaponry programs sprang up around the world. The use of bioweapons during war is banned under the 1925 Geneva Convention, and most nations have signed the 1972 Biological and Toxin Weapons Convention, which essentially bans all activities associated with the development, storage, and use of bioweapons. By the year 2000, however, between 10 and 17 countries—including Russia, Iraq, China, Iran, India, Israel, North Korea, Libya, Taiwan, and Syria—were thought to be operating covert bioweaponry programs.

Bioweapons are far cheaper and easier to produce and conceal than any other weapon of mass destruction, which makes them appealing to terrorists. The Aum Shinrikyo cult that released deadly sarin gas in Tokyo's subway in 1995 is believed to dabble in bioweaponry. Members are said to have once cruised Tokyo streets in a specially equipped van in a botched attempt to release botulinum toxin. There are reports that the same cult attempted to collect Ebola virus samples from Zaire during a 1992 outbreak of the disease.

RESPONDING TO THE THREAT
According to the United States Office of Technology Assessment, 220 pounds (90 kg) of

IN RECENT YEARS, *the military in many countries has accelerated the development of protective clothing, breathing apparatus, and detection equipment to protect troops against a range of potential biological and chemical weapons. A portable electronic detector for nerve gas is demonstrated above*

THOUSANDS *of New Yorkers wait for vaccinations against smallpox in 1947. The threat of bioweaponry has more recently prompted research into a new smallpox vaccine.*

COULD SMALLPOX RETURN?

The global eradication of smallpox was perhaps the most outstanding public health achievement of the World Health Organization (WHO) in the 20th century. Now scientists fear the disease could make a return as a weapon of mass destruction. Two WHO-controlled samples of the virus that causes smallpox remain—one in Russia, the other in the United States. There is concern, however, that covert stockpiles may exist. In 1998, a working group representing major interests in medicine, government, the military, research, public health, and emergency management, attempted "to develop consensus-based recommendations for measures to be taken by medical and public health professionals following the use of smallpox as a biological weapon against a civilian population." The group detailed its recommendations in the respected *Journal of the American Medical Association* in 1999.

tasteless, colorless, and odorless anthrax spores released over Washington, D.C., could cause between 130,000 and 3 million deaths. Victims would show symptoms long after the disease had been delivered. Science can do little to avert such a threat at present, as shown by the "anthrax letters" of 2001, but work is under way to improve defenses against bioweapons.

There are research programs developing "pathogenic neutralizing agents"—creams that could be applied to protect soldiers from agents that can penetrate the skin, such as ricin and botulinum. Some of biotechnology's most advanced methods of identifying potential new

THE CHOLERA BACTERIA, *Vibrio cholerae, could be used to deliberately contaminate drinking water supplies. Without treatment, up to 50 percent of cholera victims die.*

drugs are being applied to develop bacteria-killing products. Work is also under way on the development of rapid field tests for organisms such as plague. Airborne biological agents—the ultimate stealth weapons—are extremely difficult to detect, but new aerosol collector technology now under development may provide useful biodetector systems. The threat of bioweaponry is also accelerating some work in the development of vaccines, including research into a new smallpox vaccine.

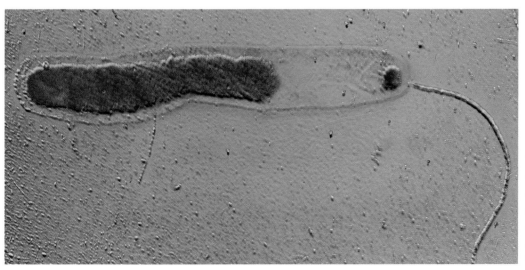

There is a theory which states that if ever anyone discovers exactly what the Universe is for and why it is here, it will instantly disappear and be replaced by something even more bizarre and inexplicable.

The Restaurant at the End of the Universe,
DOUGLAS ADAMS (1952–2001), British writer

CHAPTER SEVEN

ASTRONOMY *and* SPACE EXPLORATION

Human Exploration of Space

Spaceflight is a technology that has finally come of age—but what we will do with it remains unclear.

Human spaceflight stands at a crossroads. A generation after the Apollo Moon flights—still humanity's high-water mark—astronauts no longer venture far from home. They pilot the space shuttle, they upgrade instruments on orbiting telescopes, and they busy themselves aboard the International Space Station. The action is thrilling and photogenic, and every year it inspires thousands, if not millions, of schoolkids to dream of becoming astronauts. But as late American astronomer Carl Sagan said, "exploring space isn't about flying in low orbit while tending weightless tomatoes. It's going to other worlds."

The paradox of the International Space Station is that it represents both a direct on-ramp to the planets and a cozy halfway house we may never escape from. Which role it will play in the 21st century is still undetermined.

A HOUSE IN SPACE

Learning how to live in space is the major goal of the International Space Station. It circles the

Earth 16 times a day at an average altitude of 250 miles (400 km). From time to time, various instrument-packed modules, experiments, and other equipment are hauled up to the station and attached, Lego-like, to its structure.

Some modules are for science, others are for manufacturing and processing materials in ways that are impossible under full gravity. These experiments will contribute to the development of science and technology, and may even earn the space station driblets of money. And the space station has a definite money-maker in the idea of allowing reasonably fit customers to spend a few days, or a week, on a kind of ultimate getaway space vacation.

The space station's real tasks, however, revolve around the area of human health. On March 25, 1993, cosmonaut Sergei Krikalyov returned to Earth. He had spent 311 days— more than eight months—aboard the Russian space station, *Mir*. (His flight lasted about as long as the travel time to or from Mars.) Kri-

kalyov could barely stand when he first landed back on Earth, and it took a long time before his body readjusted properly.

Krikalyov's experience is by no means out of the ordinary. Astronaut Jerry Linenger was one of five United States astronauts who spent time aboard *Mir* between 1995 and 1997; his stint lasted for four months. A physically fit man, Linenger said that while he adjusted to weightlessness within a week of arriving at *Mir*, after his return to Earth it took his body two years to recover fully.

FIGHTING BIOLOGY

Humanity evolved on a planetary surface, and we struggle against gravity from birth until death. This simple fact poses a big barrier to human spaceflight. Exploring space generally entails prolonged periods of low or zero gravity, and our bodies don't know how to handle it.

Our circulation system is built to deliver blood and oxygen to a brain standing roughly 6 feet (2 m) above the toes, and to do it without putting too much or too little pressure on either extremity. In the absence of gravity the system pumps too much blood into the head, so that it grows puffy, while the legs and feet grow spindly. Muscles atrophy from lack of use. Even the body's immune system weakens for unknown reasons; colds become more prevalent and are harder to shake off.

ASTRONAUTS ENDURE *long, hard periods of training before going into space. Their training will include simulations of gravitational effects in 15-G centrifuges like the one at the astronaut training center in Cologne, Germany (above); they will also learn to use manned maneuvering unit flight simulators (below).*

The biggest worry, though, is bone deterioration. Every trip into space starts astronauts' bones decalcifying at an accelerated rate compared with what happens on Earth. Worse yet, the deterioration is hard to reverse. A few astronauts have suffered up to 20 percent reduction in bone mass (although 10 percent is more common). Besides a much-increased risk of osteoporosis, astronauts journeying eight months to Mars, say, might arrive there too weak to handle the physical effort of landing a spacecraft

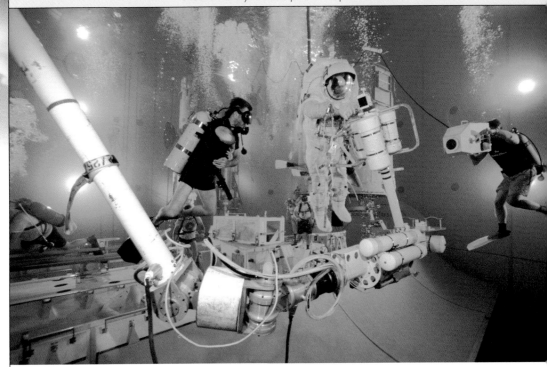

PRACTICING IN CONDITIONS *similar to those in space is a key part of astronauts' training. For instance, they rehearse repairs in a giant tank (above) because that simulates the weightless conditions they will experience in space.*

ARE WE THERE YET?

Flights to other stars so far lie firmly in the realm of science fiction, although some serious thinking has been done on proposals for propulsion. They include light sails, which might work if the spacecraft gets up to interstellar velocity while still near the Sun. Another idea is nuclear rockets fired by fusion and perhaps fed by scooped-up interstellar matter. Though far beyond today's technology, these techniques appear to violate no physical principles.

If the propulsion question looks tough but solvable, the human ones aren't so simple. Since flight times will last hundreds or thousands of years, the crew would need to be put into long hibernation. No one has any idea how to do this practicably. The alternative is a ship large enough to carry a crew about as populous as a medium-size town. Entire generations would be born and would die in space, committed to a project they had no say over and which they could not abandon. How realistic such scenarios are remains a question for the future.

safely and conducting the exploration that they were sent to do. And that's without taking into account the danger if an emergency arises that requires brute strength and quick action.

To some extent, space crews have been able to counteract the problems by using treadmills, wearing elasticized clothing that forces them to exercise, and by changes in diet. Yet the experience of Sergei Krikalyov and other astronauts on long flights is not encouraging. Much more work needs to be done before health concerns can be put to rest.

WHO'S IN CHARGE HERE?

There are also "human issues" to deal with in space. How will long-duration crews handle personality conflicts? Loss of privacy? Do we send mixed crews? Married couples? If the crew is international—a likely scenario—whose language is used? Will cultural differences produce friction? How much private time should astronauts have? Do you give the whole crew the weekend off, or should their "personal days" rotate?

Long before any Mars mission gets under way, the International Space Station must give mission planners concrete data on how to assemble the most effective crew. The experience of the Mir station has already underscored the importance of personality, and of effective decision-making on board. Equally vital are crewmembers with flexible, can-do attitudes. Ground-controllers need to keep crews on-task when the details of daily life become overwhelming, yet not become whip-cracking overlords, who might drive a crew to mutiny (as has nearly happened on several flights, both Russian and American).

Astronaut training and selection criteria may need to become more sophisticated—astronauts who are outstanding shuttle pilots may not be the best choices for commanding a flight of long duration. And how much cross-training is ideal for flight engineers and payload specialists? An important task for the International Space Station is to find answers to these questions.

GETTING REAL ABOUT SPACE

The toughest challenge facing the International Space Station may well be finding a reason to exist. There's a very real danger that without a big, transcendent goal to which society in general is committed, the space station will become simply a public-works program that feeds money to aerospace companies.

The engineers who designed and built the Mercury, Gemini, and Apollo manned space-

THE FIRST MEN *to inhabit the $60 billion International Space Station. Left to right they are American Bill Shepherd, and Russian cosmonauts Yuri Gidzenko and Sergei Krikalyov.*

craft—and the rocket boosters that launched them—focused enormous efforts on their development. Yet everyone involved viewed these intricate, demanding machines as simply stepping-stones to the Moon. The space station can help us to reach the planets. Yet unless society commits to exploring the solar system, there's little point in having a space station.

Appropriate big goals to set before us might include establishing a Moon base or undertaking a Mars expedition. Others also beckon—for example, exploring, and maybe exploiting, the mineral resources of asteroids. No other pieces of celestial real estate come closer to Earth or are so easy to reach from orbit. And economics apart, someday humankind's survival may depend on sending a crew to deflect an asteroid whose orbital path has Earth in its crosshairs.

Wernher von Braun, the developer of the giant Saturn 5 Moon rocket, was once asked what was needed to get to the Moon.

He replied, "The will to do it."

The technology to explore the solar system is at hand, and the biological knowledge to do it safely lies within our reach. But humanity still has one important decision to make, and that is whether it will spend the 21st century becoming a spacefaring civilization, or whether it will just tend a small garden while turning its back on the infinite possibilities all around.

Probes to the Planets

The search for life beyond Earth is something that focuses all

our knowledge and ignorance of the solar system, and it will

drive much planetary exploration in years to come.

The dawn of the 21st century marked both an opening and a closing in the world of space exploration. The golden age of planetary exploration had ended. Every major planet bar one, and many of the moons, had seen at least one spacecraft visit.

For an entire generation, the space frontier marched steadily outward: the Moon in 1959, Venus in 1962, Mars in 1965, Jupiter in 1973, Mercury in 1974, Saturn in 1979, Uranus in 1986, and finally Neptune in 1989. Ironically, the Cold War superpower rivalry that drove this exploration began to crumble in 1989, the year that saw the last first-visit of a planet. The collapse left distant Pluto as unfinished business, and it still awaits its first reconnaissance.

In the 21st century, the driving force behind planetary exploration is shifting. If national competition once launched rockets, probes, and people into space, what now spurs missions is largely a search for extraterrestrial life.

THE SEARCH FOR LIFE

One consequence already seems clear. The Moon will take a back seat as the new century begins. Lunar science has important problems to solve, but they are no longer at the frontier of space exploration. Mars is a very different story. This small, desert planet about half of Earth's diameter offers by far the best chance to find life, or at least evidence of its passing. The early histories of Earth and Mars ran practically in parallel as both suffered heavy impacts by asteroids, meteorites, and comets. During this violent era, life started on Earth and survived. Looking at Mars, it seems reasonable to think that life may have started there, too.

After perhaps their first billion years, Earth and Mars evolved differently. Earth's stronger gravity holds onto water and air more easily, and Earth has more internal heat to power tectonic recycling. Our planet's geo-engine is still running, but that of Mars seems to have sputtered to a halt. Mars turned cold and dry as much of its primordial atmosphere and water escaped into space, sank into the ground, or settled in the polar caps. Ultraviolet (UV) sunlight has scorched its surface rocks and dust and made them hostile to organic matter. But before Mars became a frigid desert, life might have thrived long enough to leave fossil traces in its rocks.

It was such microscopic fossil remains that scientists thought they had discovered in 1996 in the famous Mars meteorite ALH 84001 that was found in Antarctica. But after intense study, many researchers conclude that this rock bears no biological traces. Disappointing,

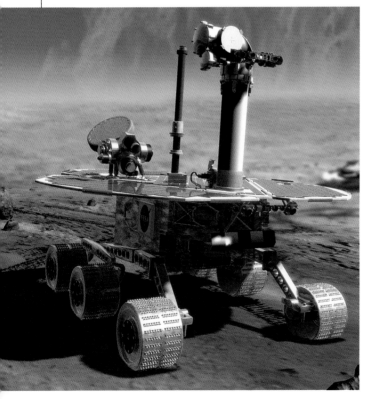

ROVING AFAR, *NASA's Mars Exploration Rover (left) will carry a panoramic camera, a microscopic imager, and other tools that will enable it to examine Mars rocks in detail. Landing is scheduled for early in 2004.*

NICKNAMED "SOLID SMOKE," *aerogel (held above by scientist Peter Tsou), is a low-density, lightweight material that can capture fast-flying comet particles and interstellar dust in space without physically altering or damaging them. Used on NASA's Stardust probe, aerogel has sampled a comet's tail and is bringing it back to Earth.*

perhaps, but it doesn't rule out the possibility of ancient Martian life. It could simply be that we haven't yet found the right rock.

MISSIONS TO MARS

The most ambitious program of Mars missions belongs to the United States's National Aeronautics and Space Administration (NASA). It will search for sites where water may have collected in the ground or flowed at the surface earlier in the planet's history. It also seeks places where groundwater may reside today. Water is the key, because scientists consider it unlikely that any life could exist without it.

NASA's Mars plans combine surveys from orbit with mobile rover landers. Once their results have told scientists enough about Mars' geology and environment, missions to bring back samples to Earth will follow.

In orbit at present are two spacecraft, the Mars Global Surveyor and the Mars Odyssey Orbiter. The Global Surveyor, launched in 1996, carries a camera that can photograph boulders as small as a few yards (meters) across. Other instruments measure Mars' gravity, topography, magnetic field, mineral composition, and dust in the atmosphere.

The Mars Odyssey Orbiter was launched in 2001. It carries a thermal imager that seeks an ancient Martian equivalent of Yellowstone Park in the United States, a place where volcanic heat and groundwater produce hot springs in which bacteria might exist now, or existed long ago. It is also hunting for dry lakebeds, where life might once have lived. Its gamma-ray spectrometer has found lots of hydrogen in the upper 3 feet (1 m) or so of soil, a telltale mark of ice or water. It can identify any salts present, and it will also track changes in the Martian polar caps.

221

The next launch window, in 2003, will see three spacecraft depart Earth for Mars, two from NASA and one from the European Space Agency. NASA's two spacecraft will place two rover landers on the Martian surface in 2004, each bigger and more capable than the Sojourner rover that accompanied Mars Path-finder in 1997. The highly successful Sojourner rolled a total distance of about 300 feet (90 m), but the new Mars Exploration Rovers will go that far each Martian day and should operate for at least 90 Martian days. Besides possessing greater endurance, they also carry five experiments each, plus a rock abrasion tool. The instruments with which they are equipped are a high-resolution panoramic camera, a microscopic imager, and spectrometers for analyzing minerals and mapping chemical elements in rocks. Together with the rock abrasion tool, the imager will function like a geologist's hand lens, giving extreme close-up views of rocks'

fine-grained appearance. The two identical rovers will go to different sites on Mars, chosen from Global Surveyor images and other data.

The European Space Agency's 2003 mission is named Mars Express, and plans call for both an orbiter and a lander. The latter is named Beagle 2, after the ship that carried naturalist Charles Darwin on his epochal world voyage. Launching in June 2003, Mars Express will arrive at its destination in December 2003, about a month before the NASA rovers.

The Mars Express orbiter carries seven instruments. These will record three-dimensional images of the surface; use ground-penetrating radar to look for deep subsurface water; study the composition of the atmosphere and how it changes and interacts with space; and map the mineral content of the surface. Beagle 2's mission is to look specifically for life.

Landing in Isidis Planitia, 10° north of the equator, Beagle 2 will stay where it lands, lacking the rovers' mobility. But it does carry a "mole" that can crawl short distances to burrow under rocks for samples. These samples have been sheltered from the Sun's damaging UV light, which hits the Martian surface

THE MARS EXPRESS MISSION *will explore Mars in 2003. After the Beagle 2 lander is launched (left), three airbags will be deployed to cushion its landing. A model (below) shows the complete Beagle 2 lander on the surface of Mars, with the solar panels extended. It will weigh about 60 lbs (30 kg).*

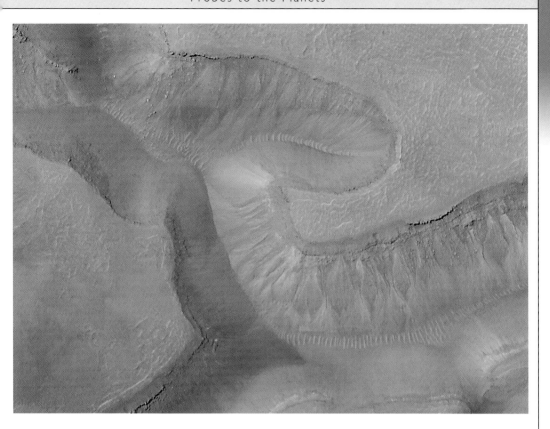

unabated. Scientists fear that UV light has destroyed organisms at the surface, but those buried in the soil or inside rocks may have survived. Getting at the latter is the job of Beagle 2's drill and grinder.

Guided by small cameras, Beagle 2's robot arm will reach for a suitable-looking rock. The grinder will remove any weathered crust, and the lander's high-powered microscope will study mineral textures and check for fossilized bacteria. Other instruments will test the rock's mineral composition and internal structure. Then Beagle 2 will drill into the rock for a core sample, which will be chemically tested for carbon. If there is life or its remains in the surface rocks, Beagle 2 should find it.

What's next? In 2005, NASA will send off the Mars Reconnaissance Orbiter, designed to span the gap between orbital views and the close-ups that landers and rovers provide. It will photograph thousands of Martian landscapes at a high enough resolution to spot rocks the size of beach balls. In 2007 NASA proposes to send a long-range mobile science laboratory to the surface. It will hunt out sites for a future sample return mission. Also possible are new small "scout" missions that may carry airborne vehicles or small landers.

The French Space Agency plans four small Netlanders as seismic and meteorologic stations, while the Italians will fly a communications orbiter to link the Netlanders and Earth.

FOLLOW THE WATER *is a theme that will drive Mars exploration, since liquid water is essential to life. High-resolution images of Mars show gullies like the one pictured above that appear to have been carved by water.*

In the following decade, NASA plans two missions to return Mars samples to Earth, launching in 2014 and 2016.

EUROPA'S OCEAN

Mars offers the best chance for extraterrestrial life, but it isn't the only place where biology might exist. Scientists believe that Europa, one of Jupiter's moons, possesses highly interesting prospects for some form of life.

Jupiter's Europa is a little smaller than our own Moon. It drew intense interest during the Galileo orbital mission at Jupiter. Earlier Voyager spacecraft had found a smooth, white, ice-covered moon, scarred with dark lines. The much sharper photos of Galileo showed that the ice was cracked, upheaved, and refrozen, with giant icebergs locked in a rigid matrix. There were few craters, implying that the surface is geologically young. From Europa's density, scientists believe that it has a rocky core and a deep global ocean of briny water or slush, topped by an icy skin. The dark lines resemble fresh-frozen openings in polar sea ice on Earth. Some regions show evidence of warming from below; mineral salts stain other features. Driven by tidal tugging from Jupiter and Europa's

223

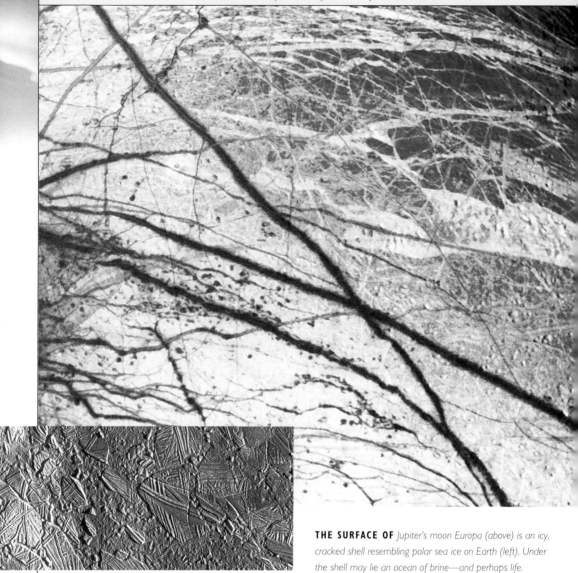

THE SURFACE OF *Jupiter's moon Europa (above) is an icy, cracked shell resembling polar sea ice on Earth (left). Under the shell may lie an ocean of brine—and perhaps life.*

sibling moons, the ice shell slides on the water (or slush), rotating at a different rate to Europa's solid body, which sets up stresses that in turn crack the ice.

There are many things that we still don't know about Europa. Is there an ocean of liquid water down there? How thick is the ice? Is the ocean kept warm by geothermal heat sources? Could life exist there? To answer at least some of these and other questions, scientists would fly a Europa orbiter. The plan, however, faces many hurdles in funding and developing the technology. After arrival at Jupiter, the orbiter would spend a year studying the planet and all its moons, after which the spacecraft would drop into orbit around Europa. There it could last about a month before damage from Jupiter's intense radiation belts puts it out of action. In that month, the orbiter must image the surface closely, map its topography, and use radar to measure the thickness of the ice. If the ocean exists and if the ice proves not too thick—two

big ifs—then a future mission might land and look for evidence of life, perhaps by melting through the ice and dropping a "hydrobot" (a submarine spacecraft), into the Europan ocean.

FURTHER MISSIONS

The search for life is the inspiration for much planetary research, but not all. Upcoming missions include ones to Sun-roasted Mercury, last visited in 1975 by Mariner 10. The soonest to go is NASA's Messenger spacecraft, which will launch in 2004 and arrive in 2008. After flying past Mercury twice to reorient its course, Messenger will spend two years orbiting the planet. The prime task is to photograph the half of Mercury unseen by Mariner. It will also study Mercury's magnetic field, its massive iron core, and search for ice in shadowed polar craters.

The small bodies of the solar system are also attracting study. A Japanese asteroid spacecraft, Muses-C, will be launched in late 2002 toward the asteroid 1998 SF36. It will arrive in 2005

OTHER MISSIONS

Spacecraft	Origin	Target	Arrival
Near-Earth Asteroid Prospector (NEAP)	USA	asteroid Nereus	2002
Smart-I	Europe	the Moon	2002
Lunar-A	Japan	the Moon	2003
Selene	Japan	the Moon	2003
Nozomi*	Japan	Mars	2003
Contour	USA	Comet Encke; two others	2003+
Cassini/Huygens*	USA	Saturn's moon Titan	2004
Stardust*	USA	Comet Wild 2	2004
Mercury orbiter	Japan	Mercury	2009
Dawn	USA	asteroids Ceres, Vesta	2010
Bepi Colombo	Europe	Mercury	2013
New Horizons	USA	Pluto	2015

* already in flight

and, after surveying the body from low orbit, it will land and take samples from three sites for return to Earth in 2007.

Asteroids are rocky debris left from the solar system's formation; comets share the same source, but they were formed much farther from the infant Sun, out where temperatures are vastly colder. A comet really becomes a comet only when it sails in close enough to the Sun—roughly as close as Jupiter—for its ices to warm into gas. As the nucleus ice evaporates (like a brick of dry ice on a hot day), the gases also carry off dust particles. Together, both gas and dust form the distinctive tails for which comets are known.

In early 2003, the European Rosetta mission heads off to Comet Wirtanen, arriving in 2011 after flying past two asteroids on the way. The mission will spend several months with Comet Wirtanen, photographing its nucleus and investigating its composition. Then, on locating a suitable site, Rosetta will place a lander on the comet's surface. Thereafter, orbiter and lander will ride Comet Wirtanen for a year, right through its closest approach to the Sun.

This mission, dramatic as it is, still looks only at a comet's surface. To look inside a comet, scientists will launch Deep Impact, a spacecraft that will fire a 770-pound (350-kg) copper projectile into the nucleus of the Comet Tempel l— and study what happens. Launch is in early 2004, and impact comes in mid-2005. The projectile will be released a day before the impact, as the main spacecraft slows down slightly and diverts its course. The collision will be photographed by imagers and spectrometers on board the main spacecraft—and also from Earth, where mission planners expect that the flash of impact should be visible in backyard telescopes.

As international cooperation replaces competition, the exploration of our solar system home has become a goal for humanity's collaborative efforts. In another sense, however, this quest is really a search for our own origins, and no one can say where it will lead us.

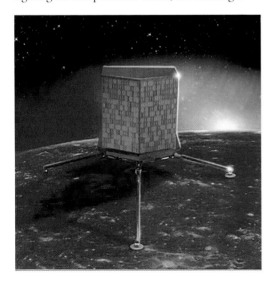

WRAPPED IN SMOG, the surface of Saturn's moon Titan is the target for the Cassini/Huygens lander (artist's impression, above), due to set down in 2004. Scientists speculate it will find craters, mountains, and perhaps lakes of hydrocarbons. The European Rosetta mission (artist's impression, left) will rendez- vous with a comet in 2011, and deposit a lander on its surface.

Colonizing the Solar System

Going into space is the inevitable next step for a species that

has never yet stopped looking over the horizon.

The urge to explore beyond familiar horizons is one of the driving forces of humanity. Our species has succeeded in colonizing virtually every ecological nook and cranny of the planet, a feat few other creatures can match, except perhaps for bacteria.

The concluding episode in this great global migration—the peopling of the world—is a lot more recent than many realize. One view holds that the process finally came to an end when canoe-borne Polynesian settlers arrived in un-inhabited New Zealand about 1,000 years ago. Another view says that it was finished as late as the 1950s, as permanent bases started to appear on the Antarctic ice cap.

To the Moon

The first step in the "peopling of the worlds" might logically be the Moon. Earth's sibling has been described as the eighth continent, because its area about equals that of Africa. Just a three-day trip away, it has already been visited by astronauts, and scientists now have a reasonably good idea of the conditions and resources that

await anyone planning to settle there. But the findings induce caution, at least with respect to the idea of a self-sufficient colony. Solar flares and coronal mass ejections from the Sun shoot off high-energy particles that are dangerous to living creatures. Earth is protected by its magnetosphere and atmosphere, but a lunar colony would have to be underground for protection. This is also necessary to shield it from the enormous day/night temperature cycle, which spans more than 450°F (230°C).

Another hurdle facing off-Earth settlements lies in their economics. Governments may fund scientific research bases if there is an overriding military or national prestige purpose. But the less compelling the national purpose, the louder the dissenting voices will become over expense, so that settlements may face pressures to be at least partly self-supporting.

Looking at the known composition of lunar rocks, it's hard to identify substances that could be mined profitably for export. And most of the Moon rocks require extensive treatment to be turned into materials that would be directly

ICE DEPOSITS *at the lunar South Pole, mapped by the*
United States' Clementine and Lunar Prospector spacecraft,
may provide a unique location for a solar-powered colony
(see artist's rendering, above). Such a base could produce
fuel, launch scout ships, and draw on local resources. A lunar
surface rover vehicle (see artist's rendering, above left) would
allow explorers to venture far from the base camp.

useful to colonists. Collecting solar energy for
microwave transmission to Earth could help a
colony earn its keep. But the expensive arrays
would have to girdle the Moon to keep power
flowing throughout the two-week lunar night.
(Also, power can be generated more cheaply on
Earth, of course, at local scales.) For the settlers
themselves, many biological necessities (carbon,
nitrogen, and phosphorus, for example) are
absent or scarce on the Moon, and would need
periodic resupply from Earth, even after agriculture is established in hydroponic farms.
Every bit of water would have to be recycled
endlessly, regardless of its source, and a certain
amount of replenishment would be necessary
to make up for the inevitable losses.

Despite these practical difficulties, the Moon
confers huge benefits for science, besides its
value as a body for geological study. For ex-

ample, the Moon's far side is nearly ideal for
locating highly sensitive radio telescopes. Permanently facing away from Earth, such sensitive instruments would be shielded from the
electromagnetic roar of human civilization by
more than 2,000 miles (3,000 km) of rock.
Today's Earth-bound radio telescopes (and
tomorrow's orbiting ones) have to snatch what
data they can from the slivers of the spectrum
that are not crowded out by broadcasting and
other commercial uses. On the Moon, giant
"ears" could listen in on the universe at any
frequency, without fear of disturbance. And the
Moon's slow rotation (28 days) gives astronomers a long time to view objects of interest.

NEW MATERIALS, *such as this heat-absorbing mesh (right),*
can help build off-Earth settlements. Years of experience with
Antarctic bases will give mission planners a headstart, but many
new, off-Earth challenges will demand new solutions.

COLLECTING LUNAR OR MARTIAN *rock samples is better executed by on-site geologists than robot spacecraft. In this artist's rendering (above), a geologist drills into the lava cliff surrounding the giant shield volcano Olympus Mons on Mars.*

Optical telescopes would also benefit from a lunar site. While the orbiting Hubble Space Telescope is an impressive instrument, it shares the difficulty of all free-flying space observatories—holding a precise aim. In space, there is no up or down, left or right. Acquiring a target, and keeping a precise lock on it while trying to collect data, is not simple when the telescope's aim is free to drift in any direction.

A lunar telescope has solid ground for its mounting, but only one-sixth as much gravity to deal with compared to Earth. Moreover, because it does not go into Earth's shadow and out again every 90 minutes (as telescopes in low-Earth orbit do), a lunar telescope escapes the frequent massive temperature changes that contribute to pointing errors. During the lunar day, an optical telescope can work as easily as at night, because without an atmosphere the sky is inky black, even at noon. Without a filtering atmosphere, lunar telescopes that are designed to detect millimeter waves, ultraviolet light, infrared radiation, X-rays, and gamma rays could all work at full efficiency.

The Moon will almost certainly be the first outpost off-Earth, but more likely as a scientific station than a full human settlement. Yet whatever the Moon's future is, a more distant place has always exerted a stronger tug on humanity: Mars.

The Invasion of Mars

In 1948 rocket engineer Wernher von Braun sat down with his Nestler slide rule and drew up a detailed plan for a manned Mars expedition. In the style of the times it was large: there were 10 spaceships, the crew numbered 70, and the entire expedition would be gone for almost three years. It used only technology existing at the time, or clearly foreseeable. Yet the cost, he estimated, would be no greater than for a small military operation in wartime.

A Mars expedition today would use a very different method, and because of the distance from Earth it would need to live much more "off the land" than is possible with the Moon. As far as a colony goes, the lunar and Martian environments differ greatly. Mars has a 24.6-hour day, a very thin carbon dioxide atmosphere, icy polar caps of water and carbon dioxide, and bracingly cool temperatures that range from 80°F (27°C) to -200°F (-130°C). Mars also appears to have near-surface groundwater in areas. Ultraviolet light strikes the surface unhindered (but it can be filtered),

MAKING SUPPLIES *would be the task of the automated "fuel factory" pictured above in this artist's conception. Such a fuel factory would potentially be sent to Mars to extract water and oxygen from the local environment. Many such missions would need to precede permanent bases on Mars.*

and solar flares have less of an effect than on the Moon since Mars lies farther from the Sun than Earth. Mars colonies could be largely located at the surface.

Currently, NASA and the European Space Agency are making ambitious efforts to survey Mars with robotic probes. Their immediate goal is to locate natural water reservoirs (either existing or former) with the idea of bringing back samples that might tell us whether Mars has life now, or ever did. The same surveys are also collecting data useful for any manned Mars mission or colony, including identifying rocks and minerals at the surface.

There's little question that technology now could support an expedition to Mars, which would precede any colony. Of course, the expedition would be smaller than the massive assault von Braun imagined. One plan would be to send a rocket to Mars containing both an Earth-return spacecraft and an automated atmosphere-processing factory, which would create methane and water using Martian air (carbon dioxide) and hydrogen brought from Earth. From the methane and water would

come necessities such as oxygen through tested chemical engineering technology.

A crew of four to six astronauts would land on Mars about two years after the rocket, using a lander-habitat spacecraft. Launched toward Mars with them would be a further unit comprising a second atmosphere-processing factory and an Earth-return vehicle. Assuming the first rocket's return vehicle is usable by the astronauts, the spare would then be set down elsewhere to begin making supplies for the next crew. The first astronauts would come back in the initial Earth-return spacecraft, leaving the scene set for the second expedition's arrival. The plan is small in scale, and flexibility is its keynote. One feature of von Braun's grand plan would stay the same, however: the 2.7-year duration of the expedition, governed by the orbits of Earth and Mars.

But what about a colony? That depends on the outcome of the first several manned expeditions. Clearly, a colony could develop from clusters of expeditions sent to one area. And once Mars flights pass the initial stages, habitat modules that had been left behind could be recycled into living quarters for a colony. The tougher problem of building a Martian economy relies largely on what minerals are available on Mars and what is needed from Earth to convert those minerals into a usable form.

MAKE A SECOND EARTH

The ultimate in colonization could be to transform Mars (or far less probably, the Moon) into a second Earth, using what might be termed large-scale planetary engineering, or terraforming. It would consist of seeding the planet with chemicals or microbes to react with the surface materials. Over time these would change and transform the planet's environment, to make it far more hospitable for humans. Imagine what Mars might be like with a thicker atmosphere, perhaps even a breathable one, with warmer temperatures, with rain, and surface water. On a world thus changed, colonies of human life would spread and thrive.

Such a project lies well beyond our present science and engineering. Moreover, many people would argue there are issues involved besides just technical practicalities. For example, some people might object that transforming Mars interferes with nature. The only reply is that all life has been altering Earth since it first appeared 4 billion years ago. Until photosynthesizing bacteria started to excrete significant quantities of oxygen around 2 billion years ago, Earth's atmosphere was toxic to a great deal of life as we know it today.

Humans, of course, have been terraforming their environment from the beginning. Environmental problems most often arise from the uninformed way this is done, rather than from the fact itself. Mars is the most Earthlike of the planets, so to take the long view, its settlement is likely. Let's do it wisely.

BEYOND MARS

The inescapable fact is that the majority of the solar system today is too hostile a place for people to visit, let alone colonize. And the stars are for future generations, those with more powerful technologies (or more patience). This century's space-faring humans will find them-

A NEAT IDEA, BUT...

In the 1970s and early 1980s, space colonies had a certain vogue. Giant cities built in space to house tens of thousands of people, space colonies would orbit the Sun like hollow, steel-and-aluminum planets. Sunlight would provide the inhabitants with electricity, and any excess electricity could be beamed via microwaves to Earth for sale; crops would be grown to provide the inhabitants with food and oxygen.

Unfortunately, the engineering questions of a space colony are just as challenging as those of a colony on Mars or the Moon, and there are no local resources to call upon. Second, while a space colony might perhaps earn a living for itself through some kind of power generation, it's still hard to envision a society that would go to the trouble of building these gigantic structures just to get away from it all. Surely living on the Moon or on Mars would be easier and more attractive.

SELF-SUSTAINING biosphere experiments like the enclosed ecosystem Biosphere 2 at Oracle, Arizona, in the United States, (below) are complex. If nothing else, the difficulties faced in creating such a habitat on Earth highlight the enormous problems that will be faced in building colonies away from Earth.

THIS IS THE DOORWAY to exploring the planets. The International Space Station pictured above and right in these artist's renderings has six laboratories on board. Europe's Columbus state-of-the-art laboratory (right and above left, front) accommodates experiments across a range of disciplines.

selves confined for the most part to the Earth-Moon-Mars zone. It is possible that future technologies—impossible to predict, almost by definition—may open up realms of the Sun's domain that are forbidden to us now. But until that day arrives, it will be the role of robot spacecraft to boldly go where we cannot.

Beyond questions of feasibility, however, some people object to the idea of humans settling on other worlds. We'll just spread all of our bad actions and bad habits to new lands, they say, infecting them with the problems and woes of humanity. It's true to say that any human colony anywhere will be a microcosm of humanity here, and will therefore inevitably have its fair share of misery and crime. True

enough—but if we have to wait until we have a better species of humans to send out to colonize other worlds, we may be waiting for all eternity.

Here are two thoughts to finish on. The human record is not entirely bad, and the very existence of a record has to be a good sign. Second, for better or worse, the desire to reach out and explore seems to be encoded in human genes, and the urge to put down new roots in new places is as old as our species. As humans we are drawn to edges and frontiers; we stagnate whenever horizons are closed.

231

Reaching into the Universe

The need to collect more light is driving an age-old quest

for ever-larger telescopes and better detectors.

In astronomy, as in many other things, size matters. Even before the telescope was invented, astronomers knew that bigger instruments were essential for learning more about the universe. Large quadrants, octants, and sextants once enabled astronomers to track planetary motions accurately enough to discover how gravity worked. Today, large telescopes allow astronomers to search for the earliest generation of galaxies, to look for extrasolar planets passing in front of their stars, or hunt for exotic ices on distant bodies at the edge of our solar system.

Early telescopes were the size of children's toys, but they soon began to grow. Giant reflector telescopes first appeared around 1790, when William Herschel built a reflector with a mirror 4 feet (1.2 m) across. He used it to study and then catalogue the faintest objects he could find. His work pushed the bounds of the

known universe substantially outward, a trend that continued as larger telescopes followed. These finally reached a plateau in the mid–20th century with the 200-inch (5-m) Hale Telescope on California's Mount Palomar.

But telescope sizes couldn't grow much bigger than that until new technologies allowed new designs. Most important among these new technologies were microcomputers, which can maintain a telescope's aim on the stars using a mounting simpler and smaller than that required by scopes such as the Hale Telescope. Also, computers can keep big optics aligned more accurately, enabling astronomers to can-

BECAUSE RADIO WAVES *have a larger wavelength than light waves, radio telescopes like the 250-foot (76-m) Lovell Telescope at Jodrell Bank in England (right) need to be much bigger than optical telescopes like the Keck on Mauna Kea in Hawaii (above), to collect the same amount of information.*

THE HALE TELESCOPE'S *massive 200-inch (5-m) mirror is cleaned (above) before its aluminum is stripped and replaced. This process is repeated every two years.*

cel the worst atmospheric distortions. This technique, called adaptive optics, analyzes the twinkling light of a single star and then deforms the telescope's light beam to exactly counter-act the distortions. This lets telescopes resolve details that would otherwise be invisible.

GLASS GIANTS

In the 21st century, a new generation of huge telescopes equipped with sophisticated instru-ments and detectors is peering skyward, driving a tremendous growth in ground-based astron-omy. Whereas the 200-inch (5-m) Hale Teles-cope was once nicknamed "the Glass Giant," it now finds itself surpassed by a dozen larger ones.

The largest optical telescopes as the 21st century opened were the twin Keck I and Keck II telescopes, each with mirrors 33 feet (10 m) across. They are situated atop Hawaii's Mauna Kea, whose 13,800-foot (4,200-m) elevation provides clear, dry vistas. The Keck telescopes' immense light grasp lets astronomers survey the shapes of galaxies born when the universe was young. They can also measure minute motions in nearby stars that signal the presence of planets. And, nearer to home, they can study icy boulders drifting in the Sun's Kuiper Belt, beyond the realm of the planets.

The Keck telescopes point the way forward, and other big telescope projects are on the march. A 34-foot (10-m) scope is being built in the Canary Islands. Several telescopes in the 26-foot (8-m) range are being linked to form interferometers, composite telescopes yielding extra-sharp resolution. Such projects include a Large Binocular Telescope in Arizona with two 28-foot (8.4-m) mirrors; the Keck telescopes working as a pair; and four 27-foot (8.2-m) scopes of the European Southern Observatory's VLT (Very Large Telescope) in Chile.

Yet ahead lie such visionary projects as the 100-foot (30-m) CELT (California Extremely Large Telescope) and a 165-foot (50-m) XLT (Extremely Large Telescope) by the Lund Observatory in Sweden. And if long-standing trends in optical telescope size continue, we can expect an optical telescope with an aperture of around 330 feet (100 m) by the end of the 21st century—or much sooner, as the Euro-pean Southern Observatory is already drafting detailed plans for it. This gigantic instrument is known, appropriately, as the OWL—the Overwhelmingly Large telescope.

ALL THE KINDS OF LIGHT

By the mid-20th century, astronomers were also exploring outside the realm of visible light, at first using radio frequencies, and then other wavelengths. Early radio telescopes were crude, and their results interested few astronomers ini-tially. But as they grew in size, sensitivity, and resolution, these new telescopes revealed a uni-verse startlingly large and violent. Astronomers were quickly drawn to the new field.

THE VERY LARGE ARRAY *of 27 radio telescopes at the National Radio Astronomy Observatory in New Mexico (above). The Arecibo Radio Telescope in Puerto Rico (left) has a dish transparent to sunlight, allowing groundcovers to flourish.*

ceived at each antenna, astronomers can see tiny details in distant objects, often finding features that no optical telescope could. Astronomers hope such views will let them catch sight of matter in the act of disappearing into a black hole, or the churning fiery gas of an expanding supernova.

The 250-foot (76-m) radio dish at England's Jodrell Bank soon showed the value of a large antenna that can point anywhere in the sky. It listened in on everything from rocket probes to quasars in deep space. Other large radio dishes include one of 300-foot (90-m) at Green Bank, West Virginia, in the United States, and one of 1,000-foot (300-m) at Arecibo in Puerto Rico.

Radio interferometers also play a large role in research. The Very Large Array in central New Mexico has 27 antennas in a Y-shaped configuration that can span 22 miles (35 km). Britain's MERLIN array links seven antennas covering a distance of over 120 miles (190 km). And the Very Long Baseline Array (VLBA) uses 10 antennas from Hawaii to Puerto Rico to gain a high-resolution view of the cosmos. The VLBA's spread of antennas reaches nearly as far as any can on Earth.

The next step, naturally, is to take radio interferometers off Earth. Accordingly, future projects will place antennas in space, tied to Earth via radio links. One such is ARISE, the name for an 82-foot (25-m) orbiting radio dish that will work with radio telescopes on the ground to produce a major increase in sharpness. By carefully comparing the signals re-

WHERE THE SPACE WINDS BLOW

The light and shadows that flicker across the bottom of a swimming pool mimic what astronomers see whenever they look through the atmosphere. Adaptive optics can cancel most of this distortion, but nothing can stop atmospheric gases from robbing information from incoming light. For this, the only cure is to place a telescope in space.

However, some infrared wavelengths remain accessible from Earth-based sites that are high and dry, such as Mauna Kea in Hawaii and the Andes Mountains in South America. The new large telescopes built in these places will work at least part-time on infrared targets. And NASA's new Stratospheric Observatory for Infrared Astronomy (SOFIA) fits a 100-inch (2.5-m) telescope to a Boeing 747 and flies it above the atmosphere's worst effects. Among the objects it studies will be cool regions of dusty gas where stars are forming. Much clearer infrared views will belong to NASA's SIRTF, the Space InfraRed Telescope Facility. This spacecraft carries a 34-inch (86-cm) mirror and will orbit the Sun at Earth's distance, but trailing it by millions of miles. SIRTF's job is to study young galaxies and stars, dust disks around stars where planets may be forming, objects hidden by dust, and some of the most distant objects in the universe. SIRTF will be followed by an ambitious Far Infrared and Sub-millimeter Telescope (FIRST), launched in 2007 by the European Space Agency. FIRST carries a 115-inch (2.9-m) mirror and will float in space 93 million miles (150 million km) farther from the Sun than Earth.

SIRTF is the last of NASA's Great Observatories, four satellite telescopes designed to cover regions of the spectrum. One of the four, the Compton Gamma Ray Observatory, has already ended its mission, but two others, the Hubble Space Telescope and the Chandra X-ray Telescope, remain in orbit. Hubble will operate until around 2010, covering all wavelengths from near-ultraviolet to near-infrared. Looking beyond Hubble, astronomers are designing the Next Generation Space Telescope (NGST), for launch around 2009. Plans are still in flux, but NGST will have a main mirror 26 feet (7.9 m) in diameter and will work primarily at infrared wavelengths. Chandra will be replaced by the Gamma-ray Large Area Space Telescope, or GLAST, planned for a mid-2005 launch. GLAST will study the nuclei of active galaxies, pulsars, black holes, and supernova remnants.

HUNTING OTHER EARTHS

Perhaps the most exciting project in the planning is what NASA calls the TPF, or Terrestrial Planet Finder. TPF will place four 115-inch (2.9-m) telescopes far out in space. By combin-

ing the signals received by each telescope as it aims at one star, astronomers will be able to cancel the glare of the star and look for pin-pricks of reflected light coming from planets the star may have. The goal is to examine some 150 stars within about 50 light-years. Many extrasolar planets have been found so far, but all have been gas-giants like Jupiter or Saturn. This time, astronomers hope to zoom in on other Earths, places that we—or somebody else—might call home.

ASTRONOMERS TODAY *use sophisticated computers to control powerful telescopes, satellites, and space probes and to then analyze the images and the data received.*

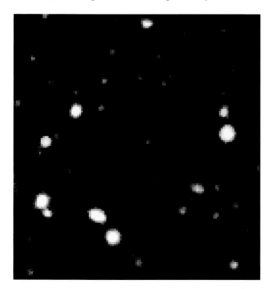

A DISTANT QUASAR *(the red point of light in the picture above) in the constellation Cetus may be vital to understanding when and how the first structures in the universe came to be.*

A SMALL TELESCOPE PROBLEM

Giant telescopes are ideal for expanding astronomy's frontier. But some astronomers view the new behemoths as a mixed blessing.

Big telescope projects place huge demands on funding, often leaving smaller telescopes—and "small" may be a mirror fully 6 feet (1.8 m) across—limping along under severely limited budgets, or facing closure. Astronomers worry that, if big telescopes dominate the field, important astrophysics questions will be ignored.

History shows examples where lengthy studies of a few stars, or broad surveys of many objects such as galaxies, have greatly advanced the science. Such research, however, requires many observing nights throughout the year—exactly the sort of proposal that gets rejected for big telescopes because the time it requires prevents access for too many other astronomers.

The solution lies in funding productive telescopes of all sizes. Yet because tight budgets are an inevitability, how this tug of war plays out may well determine the future of astronomy.

Beginnings and Endings

The search for extraterrestrial life may be the most far-fetched—and the most human—quest of all.

Sometime this century we may know. It might come as a message from space, or as a faintly heard signal plucked from the cosmic background noise. We might, of course, find it closer to home, drifting in the dark, briny sea of a gas-giant planet's moon, or as unmistakable traces locked in a rock that comes from some-place other than Earth. Whatever the exact outcome, an age-old question could finally have an answer: Is life unique to Earth?

The search for life beyond Earth is not a question that lies foremost in the minds of most astronomers and planetary scientists. In fact, to mention it in the wrong places can get research grants abruptly curtailed. But astronomy and the search for life both grow out of the same curiosity that has sent human beings to explore and inhabit the Earth from one pole to the other. So what is it that drives our curiosity about extraterrestrial life? Certainly not economic gain, as it's no road to wealth. It could be that humans are simply a gregarious species

TAKEN BY VIKING *in 1976, this aerial photograph of a Martian butte was seized upon by the paranormal community as evidence of civilization on Mars, even though such illusions are common in nature.*

and we wish to know that we are not alone. But a deeper look suggests that, in hunting for life in the universe, we are really searching for clues to our own origins. If we can find one biosphere away from Earth—one genuine "exobiology"—we will finally start to learn more about what we are.

EARLY EARTH

As a first step, it's worth examining the earliest forms of life here on Earth. Our planet formed about 4.56 billion years ago from a cloud of gassy dust that surrounded the newborn Sun. Because Earth took shape in the inner part of that cloud, the minerals building it were those that could survive warm temperatures: Silicon, iron, oxygen, and others. At Jupiter's distance and beyond, low-temperature substances like hydrogen gas and ice predominated.

Earth grew by the collision and accretion of rocky (and some icy) debris, which made our earliest geo-history extremely violent. At that time Earth resembled the heavily cratered southern highlands on the Moon. Although this might sound hopelessly hostile to any form of life, it wasn't always to be so. Along with this rocky debris came water and organic compounds from comet impacts. Volcanoes erupted water. Lakes and oceans would have exist-

PROTECTED BY ITS ATMOSPHERE, *warmed by the Sun and with plenty of water, Earth (seen from the Moon, left) is the only planet in the universe known to harbor organic life. Earth is also the only planet where water is known to exist as a liquid on the surface. And anywhere that liquid water is found on Earth today, life will be found there also.*

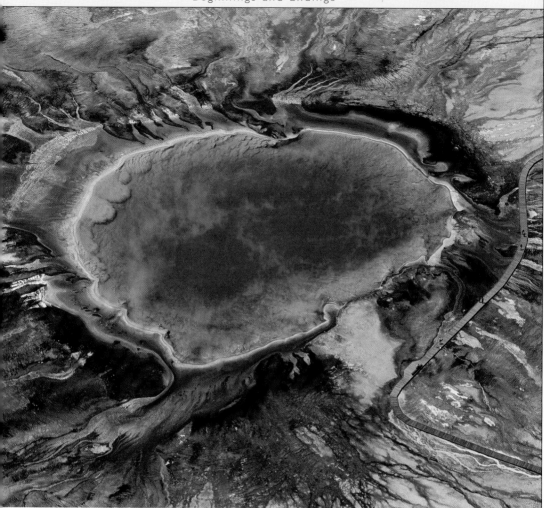

ed, even if they periodically evaporated in the heat of impacts and recondensed. Life may have started and been wiped out several times.

The oldest recognizable fossils, in 3.5 billion year old chert found in Western Australia, are of bacteria and closely resemble some bacteria of today. Chemical evidence also exists in some rocks for life at 3.9 billion years ago, close to the time when violent impacts on Earth finally slowed significantly. Whatever the exact history of that tortured era, life clearly emerged under very tough conditions. Today, life is found on Earth anywhere that liquid water is found, including environments as extreme as Antarctic lakes covered in ice, underwater volcanic vents, and deep subsurface rocks. In searching for exobiology, we need to remember that life may well be thriving in some unexpected places.

WE'RE ORDINARY

People sometimes wonder if there is something special about the chemistry on Earth that permits life. The answer is no. To the best of our knowledge, the chemical elements found here on Earth are found everywhere in the cosmos. The interstellar molecules detected by our radio

A HYDROTHERMAL VENT *in Yellowstone National Park in the United States is surrounded by brown, yellow, and green cyanobacteria. Since life evolved on Earth, it may have evolved elsewhere, too, but in looking for exobiology it is important to remember that life can exist in such unusual places.*

telescopes, the elements that form in stars and are then ejected in supernova explosions, and the basic star-stuff that drifts in giant clouds of gas and dust, all seem to have the same chemical affinities and capabilities as their local examples. Two atoms of hydrogen and one atom of oxygen will combine to make water anywhere.

It is possible that an alien life form's biochemistry may be quite different from ours. But in a universe built from the same table of chemical elements, it's reasonable to look for exobiologies based on the one element that provides the most diverse ways of combining with other elements. That element is carbon, and it forms the basis for our life on Earth, too.

Life requires carbon and water, but it also requires energy. Scientists used to think that life needed access to solar energy, either directly or indirectly. But that was before the discovery of organisms around deep-sea hydrothermal

237

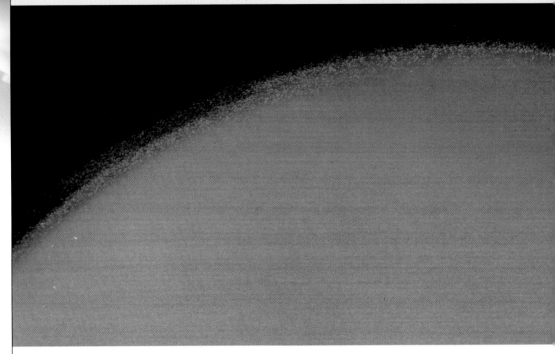

springs. These live on minerals spewed into the hot water. And bacteria have been found living in the pore spaces of rocks, at a depth of half a mile (1 km) or more. These subsist on energy extracted from the rock around them and the natural heat flow of the planet.

Terrestrial life displays certain traits, which are probably useful in looking for it elsewhere. Life uses energy and gives off waste products. It reproduces; life forms may move on their own. And they certainly go through evolution. But besides looking in the more unlikely places for traces of life, we ought to keep in mind that exobiology, if ever found, may challenge our very criteria for life.

SATURN'S *largest moon, Titan (in the computer-enhanced photograph above) is thought to be covered in organic compounds similar to those from which life originated on Earth.*

WHERE TO LOOK?

Life is adaptable, but that doesn't mean it lives everywhere. "Follow the water" is a reasonable principle. It lies behind the efforts of NASA, the United States' National Aeronautics and Space Administration, to survey Mars, the nearest likely home of exobiology, either presently existing, or a former exobiology now extinct. (The latter is much more probable.) Thus NASA has proposed a sequence of ambitious missions to search for traces of life on Mars where water has flowed or collected.

Other sites to check include Jupiter's moon Europa, which has a deep ocean of brine, and Saturn's moon Titan, whose surface is covered by organic compounds not dissimilar to those from which life probably originated on Earth.

WHERE ARE THEY?

Enrico Fermi (left), famed nuclear physicist and co-developer of the first atomic reactor in the 1940s, was skeptical of extraterrestrial life. He saw that terrestrial organisms quickly spread to occupy every possible ecological niche. Because our solar system is a relative latecomer to the universe, which existed for roughly 10 billion years before the Sun appeared, any extraterrestrial life forms have had sufficient time to colonize every galaxy thoroughly. So, he asked, where are they? His question had no answer in the 1940s, and it has no answer now.

THE SUN *provides most of our energy in the form of sunlight; indeed many plants make food just by standing in sunlight. But not all life needs solar energy. Some bacteria and organisms subsist on energy extracted from the rocks around them and from the natural heat flow of the planet.*

There is even speculation that, since the atmospheres of gas-giant planets are rich in organic compounds, they might harbor microbe-sized life that has an aerial existence, drifting on the winds and updrafts.

Why stop there? Since the mid-1990s, astronomers have discovered literally dozens of stars that have planets orbiting them. These planets are massive, presumably similar in size to Jupiter and Saturn. But astronomers expect that a great many of these planetary systems will have smaller planets as well—perhaps Earth-like ones. If any of the smaller planets happen to orbit the star at a distance where water can remain liquid, they might well have biospheres.

TALK, THEN LISTEN

Extraterrestrial microbes are one thing, but what about other intelligent species? In the 1960s radio astronomer Frank Drake wrote a now-famous equation to estimate the number of intelligent species existing in the galaxy. The equation begins with fairly straightforward factors, like the number of stars formed each year in the galaxy, and ends in sheer guesswork as to the average lifetime of a technological civilization. The simple answer is no one knows.

If other intelligent species do exist out there, then they live in the same universe that we do, governed by the same physics and the same chemistry. That would provide a link between us, whatever the differences in our biologies. And the physics that forbids faster-than-light travel can still provide humanity with a means of communication through radio signals.

But we'll need patience. Radio waves dawdle along at the same speed as light. And since interstellar distances are measured in years of travel at light-speed, we shouldn't bother opening any discussions that need an answer in less than one or two hundred years.

So where are we? Unfortunately, we're right where we began—looking out at a beautiful, star-strewn universe and trying to extrapolate from an example of one. It's a risky business, intellectually, but then we wouldn't be human if we didn't at least try.

THEY'RE ALREADY HERE

One reply to Fermi lies in the tabloids sold at supermarket checkouts: Extraterrestrials are already here. Moreover, they seem to have such insatiable curiosity when it comes to humans that they abduct us from time to time for strange—but usually nonfatal—medical experiments.

One has to wonder just how intelligent these alien beings are. They can build highly sophisticated spacecraft, yet they need to examine a great many humans in order to learn how we are made. And, it seems, they can't do it undetected.

Like the flying-saucer hysteria that gripped some believers during the Cold War, claims of being abducted by aliens tell us far more about human frailties than they do about extraterrestrial life.

The Theory of Everything

It would encapsulate all of existence and reveal the deepest nature of reality. Could we be on the verge of discovering a theory of everything?

Why? Any answer to this question always seems to lead to another "why?", a new question about a deeper level of reality. But perhaps this cycle can end. Some time in the 21st century, we may find the final "because"—a tidy theory that explains all physical phenomena. In this dreamed-of "theory of everything," every force of nature, from gravity to the forces that bind atomic nuclei, would be unified into a single superforce. The theory should also explain why we are all made of particles such as electrons and protons, and why these particles have the properties they do.

It sounds ambitious, but it is only a natural extension of ordinary science, which strives to unify. Charles Darwin's theory of evolution by natural selection, for example, can explain how the global menagerie of complicated plants and animals could have descended from a single, simple ancestor.

Physicists seek a more fundamental unification, a theory that applies not only to living things, but to all things. Isaac Newton made a start when he realized that gravity might extend throughout the universe, affecting not just people and apple trees, but everything. With this idea he could explain the orbits of the Moon and the planets in our Solar System; and we now know that Newton's theory works well for stars and distant galaxies, too. But he had only described one force of nature. Many things happen that are not driven by gravity.

UNIFYING THE FORCES

Electricity and magnetism were the next forces to be tamed. James Clerk Maxwell knew that electricity could create magnetism, and so he concluded that they must be profoundly linked. He created a theory that unified these two forces into a single force of electromagnetism. It proved a tremendously powerful idea, explaining light as an electromagnetic wave. Electromagnetic forces also drive chemical reactions, such as the combustion of a candle, and forge the chemical bonds that hold everyday matter together.

So could gravity and electromagnetism be brought together into a single, complete description of all fundamental forces? Albert Einstein spent thirty years trying to weave electromagnetism into his theory of General Relativity, which had replaced Newton's theory as the best available description of gravity. But it was a vain attempt.

For one thing, Einstein rejected quantum mechanics, which says that small particles behave in several peculiar ways, such as tending to blur out into a fuzz of probability. And two more ingredients were missing.

SIR ISAAC NEWTON *realized that gravity, the force that pulls apples from the trees, might also hold the Moon in orbit about the Earth and even reach out to the rest of the Universe. In doing so he unified what had seemed to be wildly different phenomena. His theory of universal gravitation described one of the fundamental forces of nature.*

JAMES CLERK MAXWELL (left) unified two forces of nature. He worked out how electricity and magnetism are woven together into a single force, electromagnetism. His equations describe its obvious manifestations, such as lightning, magnets, and electric circuitry; and also the less obvious. Light, he discovered, is an electromagnetic wave. This force also makes chemical bonds, so our body chemistry is ruled by electromagnetism.

Deep within every atom is a tiny ball of protons and neutrons, each made up of even tinier particles called quarks. These subatomic particles feel two other forces besides gravity and electromagnetism: the color force, which binds them together, and the weak force, which sometimes breaks them apart. Without including both the color and weak forces, no theory can be complete.

THEORIES OF ALMOST EVERYTHING

In the 1970s, American Steven Weinberg and Pakistani Abdus Salam devised a theory that unified the electromagnetic and weak forces. It was built on the idea that forces are actually carried by particles. Richard Feynman and others had already worked out how, according to quantum mechanics, electromagnetic forces are exerted by photons. So a magnet clings on to a fridge door by sending out a stream of photons to the atoms of iron nearby.

When Weinberg and Salam adapted this idea, they found that new particles called the W and the Z could carry the weak force—and when these particles were actually discovered in 1983, the theory was established. Although electromagnetism and the weak force seem very different, "electroweak" theory shows that they are actually two sides of the same coin. When matter is hot enough, as it was in the very early universe, just a trillionth of a second after the Big Bang, the two forces change their nature, and become identical. Only in the chilly modern universe do they seem different.

241

IN 1979, STEVEN WEINBERG *(far left), Abdus Salam (near left), and Sheldon Glashow were awarded a Nobel prize for their electro-weak theory, which unifies electromagnetism and the so-called weak nuclear force. The weak force generates a form of radioactivity, and it's thought may drive the huge stellar explosions called supernovas. A supernova remnant—the Cygnus loop—is shown in the picture above.*

How might other forces be brought in to the fold? The color force binds quarks together using more carrier particles called gluons, and there are several tentative "grand unified theo-ries," or GUTs, that encompass the color force, the weak force, and electromagnetism. GUTs can also, in theory, explain some of the prop-erties of the particles, such as their masses and charges—and physicists would love to know why the particles have the properties they do, because if these numbers were only slightly different, the universe would be very inhospi-table. Stars might burn out in a few years, or collapse into black holes. Matter might be highly unstable. But even a GUT isn't a theory of everything. Remember gravity? GUTs don't include it.

Gravity is different. Physicists now describe all the other forces of nature in terms of carrier particles, flitting between two pieces of matter. Yet, according to General Relativity, gravity is actually a warp in space. It is made of geom-etry. Einstein's theory seems to have success-fully predicted several bizarre phenomena, such as the space-eating monstrosities called black holes, so it can't be far wrong. But it is hard to reconcile his warped geometry with a stream of force particles.

STRINGS AND LOOPS
Nevertheless, one audacious theory may be on the brink of uniting gravity with the other forces. In superstring theory, the basic building blocks of the world are not point-like particles,

but tiny, writhing, one-dimensional strings. These strings live in a complicated hyperspace made up out of three dimensions of space and one of time, plus six more small, curled up dimensions that we cannot see. The way a string vibrates determines whether we see it as an electron or a quark or a photon. The American Ed Witten, one of the greatest string theorists, called it "21st-century physics that fell accidentally into the 20th century."

Its concepts may be bizarre, but its potential is awesome. Superstring theory demands that something like gravity exists, because there is a kind of string vibration that would be a graviton—a particle that carries gravity. Although there are different versions of string theory, Witten and his fellow physicists have shown in the past few years that all these versions are approximations to a single, shadowy construction that they call "M theory." They still don't know what M theory will be like. It might sit in a space of 11 or 12 dimensions, inhabited by strings and membranes and even more bizarre fundamental entities.

But other physicists point out that string theory has a fatal flaw. It assumes a smooth background of space and time as a playground for its strings, whereas a true theory of everything should explain where both space and time come from.

General relativity tells us that matter and energy curve space; quantum mechanics says that everything fluctuates at random, especially on small scales. Add them together, and at the smallest scales space-time should be a mess—a tangle of humps and bumps and holes, appearing and disappearing constantly. Physicist John Wheeler christened this stuff "quantum foam."

One attempt to put this idea on a firm footing is called loop quantum gravity. This theory is so fundamental that it steps outside space and time. At the most basic level, reality is a constantly changing net of mathematical relationships. Space, time, particles, gravity, and everything in existence is supposed to emerge from this abstract melee. But the theory is only in its early days and, so far, there isn't even a way to extract matter and forces from the mathematics.

Perhaps the theory of everything will be some fusion of loop quantum gravity and

strings? Or perhaps it will be some still unimagined idea? Despite this uncertainty, some optimists think we could find the final theory in the early years of the 21st century.

EVERYTHING?

On the other hand there may be no such thing as a theory of everything. Perhaps nature is more subtle than mathematics, so elusive that the best we can do is devise ever better approximations, each of them only true as far as it goes. And even if there is a final theory, it won't actually tell us everything. The world is built from layer upon layer of complexity. So, even if you know how each basic building block behaves, when you stick countless quintillions of them together to make a living cell or a droplet of water, you'll never be able to calculate from your theory where the cell will move, or even that the water is a liquid.

But the final theory holds out a different kind of promise. Physicists think that it should explain where the universe comes from—how and why it was created. It may be that the universe sprang out of nothing, they say. Or perhaps, by a curious twist of the quantum foam, it came out of its own future. Whatever really happened, a true theory of everything should tell us. That will be why.

THE LAST FORCE OF NATURE *is the strong force, which binds protons and neutrons tightly together in atomic nuclei. In an atomic bomb the strong force is overcome and fragments of each nucleus are blasted apart by electric repulsion.*

God not only plays dice. He also sometimes throws the die where it cannot be seen.

STEPHEN HAWKING (b. 1942), British physicist

INDEX *and* GLOSSARY

INDEX *and* GLOSSARY

In this combined index and glossary, **bold** page numbers indicate the main reference and *italic* page numbers indicate illustrations and photographs.

A

abortion pill 34
Accelerator-Driven Subcritical Nuclear Reactors (ADSNR) 81
actuators 126, 129
adaptive optics In astronomy, a technique that analyses the light of a single star then deforms the light beam to counteract atmospheric distortions, thus allowing the telescope to resolve details that would otherwise be invisible. 232–4
ADEOS Low-Earth orbit satellites carrying a wide range of environmental sensors. **195**
aerogel A lightweight material that can capture fast-flying comet particles and interstellar dust in space without physically damaging them. 221, *221*
aerosonde A small robotic aircraft used for meteorological investigations. 196, *196*
aging 54–7
  AGEs (advanced glycation end products) 56–7
  anti-aging diet 54–6
  NAD 55–6, *55*
agriculture The cultivation of land, production of crops, and raising of livestock.
  biodiversity 202–5
  genetically modified food *see* genetically modified food
  International Plant Genetic Resources Institute (IPGRI) 205
  land management 187
  machine controlled 62
  sustainable **72–3**, 187
AIDS *see* HIV/AIDS
air transport 146–9, *146–9*
  Airbus 146, *147*
  airships 148
  Blended Wing Body (BWB) design 146, 149
  catapults 147, *148*
  External Visibility System (XVS) 149
  military aircraft 148–9, *149*
  smart planes 147–8
  suborbital vehicles 149
Aldeman, Leonard 159
aluminium 75

Alzheimer's disease A degenerative disease of the central nervous system, characterized by premature senility. 36–9, *38*, 115
  drugs 40–1, *41*, 42, 44, 52
amniocentesis Removal of some of the amniotic fluid from the uterus, which is then used to test for chromosomal abnormalities or to determine the sex of the fetus. 32
angiogenesis The growth of new blood cells. 46, **48–9**
Angiotensin Converting Enzyme (ACE) 90
antibiotics 59 *see also* drugs
appliances, domestic 67, *67*
aquaculture Cultivation of the resources of the sea or inland waters, as opposed to their exploitation. 16, 180–1, *181*
archaeology The study of past cultures through their surviving relics. **96–7**, *96–7*
architecture, green Building based on the principle of sustainability. **82–5**, *82–5*
  bone-based 105
  New Urbanists 85
Arecibo Telescope 234, *234*
ARISE 234
artificial intelligence A very broad field of computer science that aims to enable computers to perform tasks that otherwise require human intelligence. **168–71**, *168–71*
  agent-to-agent communication 170
  artificial brain 170–1
  weather watching 197
asteroids Small rocky objects with a diameter of less than 600 miles (1,000 km) that orbit the Sun.
  spaceflights to 224–5
  tracking 201, 210–11
astronauts
  Gidzenko, Yuri 15, *219*
  health problems 216–18
  Krikalev, Sergei 15, 216–17, *219*
  Shepherd, Bill 15, *219*
  social issues in space 218–19
  training 217–18, *217–18*
  weightlessness, problems with 216–18
  see also spaceflight

astronomy The study of the celestial bodies, such as stars, planets, moons, comets and asteroids.
  asteroid and meteor tracking 201, 210–11, 224–5
  Cetus constellation *235*
  Earth *236*
    origins 236–7
  Europa
    exploration of 223–4, *224*
  life in space 15, 77, **216–19**, *216–19*
    colonization 226–31, *226–31*
    search for 220–5, **236–9**, *236–9*
  Mars
    colonization **228–30**, *228–30*
    exploration 220–3, *220–3*
  Mercury
    exploration 220–3, *220–3*
  Moon
    colonization **226–8**, *226–8*
    landing on 12, *13*, 216, 220
    mining 77, 226
  probes **220–5**, *220–5*
  satellites 195–7, *195*, *197*, 200
  space elevator 152–3, *152*
  space junk 81, *81*
  supernova remnant *242*
  telescopes *see* telescopes
  water, search for 221, 229, 238–9
  see also spaceflight
asymmetric warfare The sabotage of supposedly secure computer systems by computer hackers, leading to, for example, the disruption of communications or the theft or alteration of military intelligence. 175–6
augmented reality The superimposition of computer-generated images on one's view of the physical world, usually through specialized goggles. 166–7
automated beamcarriage transport (ABT) Form of transportation comprising small, driverless, computer-controlled vehicles suspended beneath a network of beams. 141, *141*
automobiles 134–7, *134–7*
  alternative fuels 135–6, *135*, *136*
  hybrid 135
  intelligent 136–7
  solar powered *134*, 135

B

bacterial mining A low-energy method of mining, using bacteria to leach minerals from ore. **75–6**
  Mt Scholl Mine 75–6, *76*

CONTRIBUTORS

Barry Anderson has a degree with honors in mechanical engineering from the University of New South Wales, Sydney, Australia. His professional engineering career included 25 years in the automotive, aircraft, and defence equipment industries, primarily responsible for product testing, prototype manufacture, new product design, product planning, and product compliance with legislative requirements. His 15 years as a consultant have continued his automotive focus by providing product design and management guidance to manufacturers, service failure analysis and product recall recommendations, and policy and procedure advice to government departments.

Stephen Battersby's first job was watching lightbulbs age at the National Physical Laboratory in Teddington, England. After reading physics at Oxford and astrophysics at Imperial College, London, he worked on the News & Views section of *Nature* magazine. He is now Deputy Features Editor for *New Scientist.*

Laurie Beckelman is an award-winning medical writer. She has published 11 books for teens, including "Hotline," a series on managing difficult emotions. Her books for younger readers include *The Facts About Alzheimer's Disease* and *The Facts About Transplants*, both of which were selected as outstanding science books for children by the National Science Teachers Association in the United States.

Marcela Bilek was born in Prague, Czech Republic. She holds a B.Sc. with honors from the University of Sydney, Australia, and a Ph.D. from the University of Cambridge, U.K. In November 2000, she was appointed Professor of Applied Physics at the University of Sydney. Her research interests are in plasma processing and fabrication of materials and renewable energy.

Michael Brooks, currently a features editor at *New Scientist* magazine in London, has written on science and its implications for a number of newspapers and magazines, including the *Guardian*, the *Independent*, and *Playboy.*

Jenny Brown is a freelance journalist who, in a long career, has worked extensively across all media. She is currently a regular columnist for the *Melbourne Age* newspaper, has written three books on health, and has produced a social history documentary for the Channel Nine Network in Australia. She resides in Melbourne, Australia.

Bruce Buckley worked in Saudi Arabia from 1987–91 with the national Meteorological and Environmental Protection Administration. He has held senior positions with the Australian Bureau of Meteorology's New South Wales office, where his duties spanned meteorological observations, engineering, and computing support for bureau activities throughout the region. He was formerly Chief Meteorologist with Australian television's Weather Channel, and is currently a senior meteorologist in the Perth, Western Australia, office of the Bureau of Meteorology.

Robert Burnham is a science writer specializing in astronomy and earth science. A former editor-in-chief of *Astronomy* magazine, he is the author of *Comet Hale-Bopp: Find and Enjoy the Great Comet, Great Comets,* and *The Star Book*, and the co-author of Reader's Digest Explores *Astronomy.* Robert has been an amateur astronomer since the 1950s, mainly observing the Moon and planets with his backyard telescope. He also enjoys following developments in cosmology.

Alf Conlon is a journalist and broadcaster based in Sydney, Australia. He has a B.Sc. from the University of New South Wales, Sydney, where he majored in philosophy and cognitive science. He has won awards for online work on the Australian Broadcasting Corporation's science website, The Lab. When he's not hosting panel discussions at "Science in the Pub," you can find him welding together robots in his garage.

Ben Crystall is Technology Feature Editor at *New Scientist* magazine in London, where he has written and edited since 1996. Prior to this, he spent seven years as a researcher at Imperial College, London, where he worked for a Ph.D. in photochemistry, and he has also enjoyed brief spells as a gin tester, deck chair attendant, and asbestos stripper.

Leigh Dayton is a California-born writer and broadcaster, specializing in science, technology, environment, and medicine. She was educated at the University of California and the University of Alberta. After a successful writing and broadcasting career in North America, Leigh moved to Australia in 1990. Since then she has worked as the Sydney Correspondent for Britain's *New Scientist*, the Pacific Rim Editor of America's *Omni* magazine, and as Science Writer at the *Sydney Morning Herald* newspaper. Leigh has worked extensively in radio and TV for commercial networks as well as the Australian Broadcasting Corporation. She has won five prestigious prizes for her work, most recently the 2001 Eureka Prize for science journalism.

Karen McGhee, B.Sc., is a Sydney-based freelance journalist whose work has focused on the areas of science and the environment for more than a decade. She has written for newspapers, including the *Sydney Morning Herald*, magazines ranging from *Time* and *Australian Geographic* to *International Wildlife*, science reference books and television documentaries for both Australian and international audiences.

David McKenzie holds the position of Professor in Materials Physics (Personal Chair) in the School of Physics, University of Sydney, Australia. He has over 250 publications in scientific literature and four patents that have been developed commercially. His research interests are the development of new materials using plasma techniques and computer simulation.

Martijn de Sterke teaches physics at the University of Sydney, Australia. His area of interest is optics and optical telecommunications. His research is aimed at developing some of the theory that underpins the novel devices that fuel the evolution of telecommunications. In 1999, he won the Australian Academy of Science's Pawsey Medal.

Julian Malnic is a Sydney-based geologist who has worked throughout Australia. He has studied, photographed, and written about mineral exploration and mines around the world and is now active as an entrepreneur exploring for minerals.

Natasha Mitchell is as a science reporter and producer with the Australian Broadcasting Corporation. She has worked in radio, television, and online producing reports and features on health and environment issues ranging from the history of blood and medicinal cannabis to computer recycling and sandmining. In a past life she trained as an engineer.

Graham Phillips has a Ph.D. in astrophysics. He is a science reporter/communicator and worked with the Australian Broadcasting Corporation's television science unit for five years: three years as a reporter on the science program *Quantum* and two years as a reporter and host on the technology program *Hot Chips*. He makes regular appearances on television and radio, and has had science columns in several newspapers, including one in the Sydney *Sunday Telegraph* for the last 15 years. Graham has written three popular science books: *The Secrets of Science*, *The Secrets of Science II*, and *The Missing Universe*.

Wilson da Silva is a science journalist and editor whose work has appeared in newspapers, magazines, and television. A foreign correspondent with Reuters for many years, he has been a journalist on *The Age* and *The Sydney Morning Herald* newspapers in Australia, a correspondent for the British magazine *New Scientist*, and a reporter and producer on the popular Australian science series *Quantum* on ABC Television. Wilson served as the managing editor of the magazines *Newton* and *21C*, as well as being the founding editor of the international quarterly *Science Spectra*. The winner of nine journalism awards, he has also written and produced two award-winning documentaries, *The Diplomat* and *Passing the Bug*. He is president of the Australian Science Communicators, a board member of the Australian Society of Authors, and a former president of The Australian Museum Society.

Abbie Thomas was most recently the editor of *Newton* graphic science magazine. She has been a science writer for 15 years specializing in health and natural sciences. She is a regular contributor to the Australian Broadcasting Corporation's Science web site and national science radio programs, with skills across both print and electronic media. Abbie has also written for *New Scientist* magazine. Her honors degree was in fish parasitology, and she still has a soft spot for all things without a backbone.

Tim Thwaites is a freelance science writer who contributes to magazines in Australia, Europe, and North America and works for universities, research institutions, and private companies. On radio he does regular science spots for the Australian Broadcasting Corporation and commercial stations in Australia and New Zealand. He particularly enjoys talking to enthusiastic researchers, children, and small furry animals, not necessarily in that order.

Paul Willis, Ph.D. is a paleontologist who has been associated with numerous important fossil excavations. His enjoyment for communicating science to popular audiences has involved him in touring Australian elementary schools with a life-size inflatable *Tyrannosaurus rex* as his lecture companion. He is currently a science reporter with the Australian Broadcasting Corporation. He lives in Sydney, Australia.

CAPTIONS

Page 1: DNA tunnelling micrograph
Page 2: Close-up of optical fiber strands against red background
Page 3: Micrograph of brain cells from complex mammal brain
Pages 4–5: Streamer chamber photo of particle tracks from high-energy collision
Pages 6–7: The dome of the Palomar Observatory on Mount Palomar, California, appears transparent in this time exposure, showing the 200-inch (5-m) Hale Telescope inside
Pages 8–9: Iris of an eye
Pages 10–11: Three-Mile Island nuclear reactor
Pages: 18–19: Micrograph of secondary oocyte (egg) for in vitro fertilisation

Pages 60–61: A field is split evenly between yellow canola and green wheat
Pages 98–99: Superconductor experiment
Pages 132–133: Time-delayed photograph of blurred motion of traffic on highway
Pages 154–155: High-speed robot experiment
Pages 178–179: Flower reflected in water drops
Pages: 214–215: View from the space shuttle *Endeavour* of the Aurora Australis over the Antarctic region. Thrusters fire on the shuttle as it maneuvers to capture a small satellite
Pages 244–245: Man carrying sign stating "Y3K is Coming, Prepare Now!"

ACKNOWLEDGMENTS

Marney Richarson and Peta Gorman (editorial assistance); John Bull (banding and cover); Angela Handley (proofreading); Puddingburn Publishing Services (index); Dr John O'Byrne (editorial consultant).

PHOTOGRAPH AND ILLUSTRATION CREDITS

(t = top; b = bottom; c = center; l = left; r = right; ba = banding; i = inset, FC = front cover, Ff = front flap, BC = back cover, Bf = back flap)
AFP = Agence France Presse
AGE = Age Fotostock
BA = Bridgeman Art Library
CB = Corbis Images
ESA = European Space Agency
HG - Hulton Getty
IB = Image Bank
NASA = National Aeronautic and Space Administration
PR = Photo Researchers Inc.
RPA = Reuters Press Agency
SPL = Science Photo Library
TS = Tony Stone

Photograph credits
1 photolibrary.com 2 Getty/TS
3, 4–5 photolibrary.com/SPL
6–7 Australian Picture Library/CB
8–9 photolibrary.com/PR
10–11 Australian Picture Library/ Peter Menzel
12 Australian Picture Library/CB
13 Australian Picture Library/ Westlight
14t NASA; b Getty Images/IB
15 AFP 16t photolibrary.com/SPL; b Australian Picture Library
17 Getty Images/IB
18–19 photolibrary.com/SPL
19i Australian Picture Library/CB
20–59ba photolibrary.com/SPL
20 Getty Images/TS
21t photolibrary.com/SPL; b photolibrary.com
22t photolibrary.com; b Weldon Owen
23t Australian Picture Library/Corbis; b photolibrary.com
24 photolibrary.com/SPL

25 Australian Picture Library/CB
26 photolibrary.com/PR
27t Australian Picture Library/CB
27b photolibrary.com/SPL
28t, 28b Australian Picture Library/CB
29t photolibrary.com/BA
29b, 30, 31 Australian Picture Library/CB 32 photolibrary.com
33t photolibrary.com/SPL; b photolibrary.com/PR
34 Australian Picture Library/CB
35 photolibrary.com/SPL
36 Getty Images/TS
37, 38t photolibrary.com/PR; b photolibrary.com/SPL
39 photolibrary.com/PR
40, 41l photolibrary.com/SPL; r Getty Images/Photo Disc
42 photolibrary.com/SPL
43, 44 Australian Picture Library/CB
45, 46 photolibrary.com/SPL
47t, 47b Australian Picture Library/CB
48t RPA; b AFP 49t RPA
49b, 50, 51t Australian Picture Library/CB
51b RPA; t Australian Picture Library
52b, 53 Weldon Owen
54 photolibrary.com/AGE
55 photolibrary.com/SPL
56, 57t Australian Picture Library/CB
57b Department of Cell Biology, University of Texas Southwestern Medical Center
58l photolibrary.com/SPL; r Australian Picture Library/CB
59 photolibrary.com/SPL
60–61 Australian Picture Library/CB
61i Corel
62–97ba, 62, 63 Australian Picture Library/CB

64 Australian Picture Library/Peter Menzel
65t Getty Images/IB; b photolibrary.com
66 Australian Picture Library/CB
67t RPA; b Getty Images/FPG
68, 69t, 69b Australian Picture Library/CB
70t AFP; b RPA
71, 72, 73t, 73b, 74, 75 Australian Picture Library/CB
76t, 76b, 77t Nautilus Minerals Corporation Ltd
77b photolibrary.com/SPL
78, 79 Australian Picture Library/CB
80 Weldon Owen
80–81b Australian Picture Library/CB
81t Reprinted with permission of MIT Lincoln Laboratory, Lexington, Massachusetts
82 Getty Images/FPG
83 Australian Picture Library/CB
84t Australian Picture Library; b Australian Picture Library/CB
85t Civano Corporation; b Getty Images/TS
86 Australian Picture Library/CB
87l photolibrary.com/Alex Bartel/ SPL; r Australian Picture Library/CB
88t Fraunhofer Institute for Solar Energy Systems; b Manhattan Scientifics, Inc.
89 Laura Stern, U.S. Geological Survey
90 Getty Images/IB
91 Getty Images/Allsports
92–93b Australian Picture Library/CB
93t Independent Media Center (www.indymedia.org)
94 photolibrary.com/SPL
95t Getty Images/TS
95b, 96t photolibrary.com/SPL